CW01278836

Berries

Acknowledgments

To my grandparents, for passing on their gardening genes

To my parents, for their insatiable interest
in new flavors and good food

To the people who have taught me so well,
and who have encouraged me to discover new berries

To all the kind people who support and inspire me.
It is because of you that I do what I do with plants
and seek to share their joyous nature!

To my dedicated editor, Julia Genazino,
for constructive discussions and great teamwork.

healthy & tasty

BERRIES

Garden and Countryside Delights

Christian Havenith

Contents

Foreword

As children, we spent a lot of time during the school holidays with my grandparents in the Eifel region of Germany, picking berries as we explored the area. Wild strawberries and bilberries were the seasonal highlights—no bush was safe when my cousins and I were around. We loaded our baskets with the fruit. Back home, our blue fingers and tongues always betrayed us—there was no fooling grandma as to why the harvest had shrunk so much on the way home!

During my later training as a horticulturalist, I got to know and appreciate the full spectrum of indigenous berries. Our experimental enterprise worked with wild fruit for professional growers. Here I was able to grow, harvest, and, more especially, savor all these hitherto unknown berries. As a young apprentice, this opened up a whole new world of flavors and delights! I was able to put chokeberries, rowanberries, medlars, and flowering quince through their paces in a quest for flavor. Many of these "innovations" way exceeded my expectations: the berry plants turned out to be extraordinarily robust, while their fruit tasted heavenly. Many of them have found their way into my various gardens. My current diversity nursery in Wassenach, Germany is open to visitors—as well as viewing, smelling, touching, and sampling the produce, they can order particular varieties for their own cultivation. I have a tasty berry available for every kind of garden, a vitamin blast for every little corner—even curious balcony gardeners seem to find something to delight them.

I also offer herb walks (sometimes in my Celtic-Druid costume) and cookery courses. Wild berries play a key role in all of this, of course, but more especially in the fall. With a glass of elderberry punch in the hand, every grove and hedgerow becomes a magical meeting point for man and nature, where the stories, and histories, of wild herbs and berries come to life.

Introduction

What exactly are "berries"?

The bright colors and diverse shapes of berries are all part of their attempts to distribute their seeds as widely as possible. This is because berry seeds have not developed wings or similar structures for distribution, so are dependent on animals as a means of transport. But animals are not very reliable when it comes to simply carrying the seeds around with them. Hence, the plants have packaged their potential descendants in an appealing way—an eye-catching, tasty casing. Red berries really stand out in a hedge, for example, and are noticed immediately by birds. Yellow fruit also attracts the attention of man and beast, while blue berries contrast with their surrounding foliage. It was this traffic-light color principle that attracted the attention of our ancestors in the primeval forest. Anything that stood out against the green leaf canopy was worth a closer look. As soon as the potential seed couriers examined the berries in more detail, they were entranced by the intense aroma of the fruit. The sweet fragrances are enticing—and the berries turn out to be a tasty foodstuff. And so a win-win relationship developed: nourishment for man and beast, while the plant becomes "mobile" and is able to continue its reproductive cycle.

From a botanical perspective

Fruits otherwise referred to as "berries" are often not fruit at all, in terms of their botanical definitions. This book, however, covers berries that may not always be scientifically designated

as such, but which are generally referred to as berries. In brief, the correct botanical terms are as follows:
Berries are formed by the flesh of the fruit growing over the seed(s) or stone, entirely covering them. In many cases, these are fruit with a single, woody pit (called a stone), containing the seeds: cherries, peaches, plums, and apricots are all examples of

"stone fruit," or "drupes." The thin, outer skin is called the epicarp; the fleshy fruit layer underneath is called the mesocarp; and the stone is called the endocarp.
The raspberry and blackberry, or bramble berry, are examples of aggregate fruit: a single berry that comprises several "drupelets," each with their own seed.
In the case of the strawberry, however, the seeds are found on the outside of the berry, embedded in the flesh. It is known botanically as an "accessory fruit," as it develops not just from the ovary, but from other parts of the plant as well.
Chokeberries, medlars, and flowering quince, on the other hand, are all examples of pomes. Their small seeds, known as "pips," are held together in a woody core surrounded by the flesh of the fruit.
A somewhat special case is the juniper berry, also known as a box huckleberry: strictly speaking, its "berries" are the nonwooden cones of a conifer.

The common feature among the berries presented here is that they have a soft flesh, are mostly spherical in shape, and are smaller than an apple. In the stricter botanical sense, these also include the "real" berries, as well as examples of the other fruit types described above.

The berries in this book

Further criteria for the choice of berries presented here are that the fruit has a good flavor, an attractive scent, and a high nutritional value. In addition, they can all be prepared in a variety of ways. Many of them have a long tradition of use from prehistoric times, so are familiar and popular. Some are less well known, but highly promising. The plants profiled here are also very robust, require little maintenance, and yield a good harvest. This means that there is something here for everyone—something enticing for berry fans, something tasty for cooking fans, reliable plants for self-sufficiency fans, compact varieties for balcony gardeners, and varieties with a guarantee of success for children and beginners. Young families and "lazy" gardeners will also find low-maintenance plant delights featured here that require little space and are very attractive to look at.

Berries in history

Berries have accompanied humans on their evolutionary journey since primeval times, from the Stone Age through the Holy Roman Empire to the modern age—across all countries and cultures. For most of the history of mankind, blackberries, strawberries, blueberries, and the like were gathered in the wild from hedges, forests, and meadows. Hunters and gatherers were a common sight, right up until the Middle Ages, often snacking on the fruit as they journeyed along. At harvest time, whole families would go "berrying" together, in order to gather supplies for the winter. This foraging for wild food was in fact subject to legal rules and regulations, so essential was the role played by wild fruit in the human diet. Ultimately, people began to grow berry plants in their own gardens, in order to ensure supplies of this vital source of nutrition. Here, their harvest was protected from other "plunderers." For many centuries, our ancestors depended on their fruit and vegetable gardens for an essential source of food. At first, home-grown berries were limited to the varieties available from the immediate surroundings. Over time, however, increasing interest and expertise in growing these plants prompted the collection of the largest and loveliest berry bushes in particular, which people then learned to propagate through cutting and grafting.

That was the beginning of a garden culture—for both subsistence and pleasure—described from the 15th century in European paterfamilias literature. This literary genre refers to books and handbooks providing advice on best practice in household and agricultural management, with the texts generally aimed at educated landowners. These books were very successful with their diverse spectrum of advice and recipes and between the 16th and 18th centuries they developed to form the basis of modern day home economics books. As wedding gifts, they proved a valuable resource for young families. The efforts of preachers from various faith traditions wanting to improve the lives of the rural poor through self-sufficiency education contributed to the dissemination of the necessary knowledge. Pomology, the renowned branch of horticulture that focuses on the cultivation of new fruit varieties by aristocratic landowners, also dealt intensively with berries. Hundreds of new plant varieties came onto the market via this route—and wild plants gradually became the modern, high-yield, cultivated plants we know today.

Being easy to cultivate, berries soon became a "wonder weapon" in improving the basic nutrition of the entire population. In times of crisis, such as during the Second World War, the planting of berries was much encouraged. With the establishment of community gardens or allotments in the 19th century, the focus was very much on nutrition through the growth of high-yield plants, as well as on opportunities for outdoor recreation. Rental costs were kept deliberately low, in order to make the gardens accessible to the poorer classes, in the drive to alleviate hunger.

It became clear, especially in Europe during and after the two World Wars, just how essential provisions from people's own gardens were. Not only was space in the countryside utilized, urban lawn areas were also planted with berries. Preserving skills reached their height, while cellars were stacked full of fruit in jars. Culinary delicacies as well as food staples, all came from the home kitchen—financial hardship and prolonged rationing stimulated an unforeseen creativity when it came to self-sufficiency.

Just how deeply the deprivations of war had affected people was evident immediately afterwards and in the decades that followed. Although foodstuffs gradually became available again post-war, people still continued to preserve produce, ensuring that their larders remained full. Not only were the many berry plants that had been planted not uprooted, there was a significant increase in the number being grown. Knowledge about their cultivation had spread, and nurseries were once again able to make sufficient quantities of the desired varieties available. In recent decades, we have become used to the year-round availability of a wide range of fruit and vegetables for purchase. In addition, ready-made products and convenience foods from all over the world find their way into the pans and microwaves of the modern kitchen. At the same time, however, the use of fresh, regional produce according to season is again becoming significant, due to an increased awareness of healthy eating. A burgeoning interest in gardens, outdoor activities, and natural foodstuffs is leading to a revival in home-grown vegetables and fruit. Not only in country areas, but also in towns and courtyards, and on balconies and terraces—we can see the development of modern gardening trends everywhere, including guerrilla gardening and the lazy bed method of growing produce in your own backyard. And it is not surprising that berry plants, with their pretty, fragrant blossoms and tasty, enticing fruit are at the forefront of these trends!

The diversity of traditional use

In earlier times, berry plants were used in a great variety of ways. In the first instance, they supplied people with vital nutrients. Their fruit, leaves, blossoms, bark, and roots also provided key ingredients for the home medicine chest. The elder, for example, was once commonly grown by farmers due to its numerous active ingredients and their everyday uses. For example, syrup made from elder blossom was found to be refreshing and soothing in the summer, and is enjoying a revival today. In the winter, hot elderberry juice was used as an immune-boosting medicine against fever, pain, infections, and viruses. The bark was used to counter vomiting and to stimulate the metabolism. The leaves, on the other hand, were used to combat bedbugs, and also provided the effective basis for an insecticide spray. Placed in tunnels dug by moles and voles, elder leaves remain a means for today's organic gardeners to drive off such animals quickly and without the use of poison.
A further example is the thorny barberry, used to protect property from intruders, both human and animal. Barberry juice is rich in Vitamin C and was important in ensuring that people remained healthy through the winter. The berries were traditionally used as a remedy for a variety of complaints, while berberine, the active ingredient in the roots and wood, was used as a yellow dye for wool, linen, silk, and leather. The hard wood was used for woodturning. The slightly toxic seeds were used as an emetic (as such, they should be removed from any preparations using the fruit; see page 128). Berry plants, with their numerous practical applications, were therefore of great use to our forefathers; but it was principally for health reasons that these natural remedies from the hedgerows were vital (see also page 24).

The berry harvest

Fresh berries taste simply wonderful—sweet or sour and juicy—and are packed full of different flavors and healthy ingredients. Harvesting at the right time is key, for both full-bodied flavor and optimal health benefits. Most berries cease ripening once they are picked and so, for both the garden and when shopping for produce, it is worth developing an eye for ripeness. The color of the fruit is one indicator: it should be intense, and fully developed. Strawberries, raspberries, and red currants, for example, should be deep red, while elderberries, blueberries and chokeberries range from blue-black through to dark violet. A further indicator is that ripe berries come away from their stalks easily, and without force; raspberries, too, come away from their central cone without resistance when fully ripe. After the eye and touch test comes the nose—the characteristic, intensive fragrance is a good indicator of ripeness.

In the garden, it helps to observe the behavior of fellow predators: once the sugar content has developed fully, birds and wasps are quick to enjoy the ripe fruit. This is when it is time for you to start picking, and also to consider protecting the remaining berries with approved anti-bird netting. Fully ripe berries are usually soft to the touch, so you will need to handle them carefully, otherwise you might squash or burst them. With thorny bushes, it is advisable to cut off the fruit-bearing branches completely. Depending on the variety, this should only be done every two years, so as to give the plant a chance to regenerate. If you freeze the twigs for a day, the berries can be simply knocked off before being sorted. This makes harvesting and processing much easier.

I do not advise the use of berry combs, because they not only pull off the ripe berries, but also the unripe ones, as well as leaves, blossoms, and buds. They damage the plant and compromise the next year's harvest; consequently, their use is officially prohibited for the harvesting of blueberries in the wild.

Have no fear of *Echinococcus multilocularis*

Gathering berries in the wild is great fun for the whole family. It is a pleasure that is often compromised, however, by fear of infection through *Echinococcus multilocularis*. This tapeworm parasite is transmitted through contact with the droppings or hairs of infected animals. Carriers include fox, mice, and other rodents. According to recent research, however, infection is primarily via pets in the home, such as dogs and cats. Wild fruit

can, of course, potentially harbor tapeworm eggs, but the actual number of infestations caused in this way is insignificant. In principle, the higher the berries hang, the more unlikely the risk of contact. Washing the fruit thoroughly also reduces the risk considerably. If you want to be absolutely sure, heat the fruit to over 125 °F (60 °C). Freezing the fruit is ineffective, however. I personally gather all kinds of edible wild plants in the woods and meadows and wash them thoroughly under running water when I get home.

Preparing and preserving berries

Jams, compotes, and similar

The most common means of preserving fresh berries for the winter is to process them into jams and jellies. In this book, you will find flavorful basic recipes for these, as well as popular fruit juices and syrups that can be easily modified according to your choice of fruit. All these berries are also suitable for freezing. Simply lay them out next to one another on a tray, and freeze. You can then transfer them to a suitable freezer container. Freezing them like this prevents the berries from sticking to one another.

Another basic and straightforward means of preserving is compote, in which berries or pieces of fruit are cooked briefly in a little water until they have reached the desired consistency. Sugar is added for flavor and shelf life and you can play around with the desired quantity in each case. Flavorings such as vanilla, cinnamon, nutmeg, or lemon balm can be added according to taste, while additional apple pectin or cornstarch achieves a viscous consistency if required. Bottled in screw top jars while boiling hot, compote will keep for about a year.

Fruit purée

Another good basis for preserving berries is fruit purée. It constitutes the basic ingredient for a whole variety of berry dishes and is especially suitable for delicacies with differing shelf lives, and for processing wild fruit.

To make fruit purée, I first of all remove any leaves, stalks, and flower remnants so that the flavor is not compromised by bitter substances. After cleaning the berries, I then cook them with a little water, depending on the firmness of the fruit flesh, until they are soft enough. The berries should always be heated for as short a time as possible, and at a low temperature, so as to retain as many of the valuable nutrients as possible. I then purée the berries and strain them through a fine mesh sieve. This removes hard skins, cores, and seeds. In the case of barberries, elderberries, and mahonia berries, in particular, sieving out the seeds is important as they contain inedible or slightly toxic substances. Recipes using blackberries or raspberries are also usually more pleasant to eat without the seed.

The resultant fruit purée can then be made into jam, jelly, or a fruit coulis. Many berries naturally contain pectin, making them suitable as a natural gelling agent. As they also contain aromatic fruit acids, they mix well with other fruit and can be prepared together with them. A fruit purée, whether pure or in combination, can also be used to make juice, wine, and syrup. Mixed with sparkling water for children or with sparkling wine for adults, berry juices and syrups are a quick and easy way to make flavorsome and refreshing summer drinks.

Adding sugar to the fruit purée and then drying it produces a snack known as fruit leather that is especially popular with children (see Sea buckthorn leather on page 125). This delicious sweet treat is an extra source of vitamins, especially in the winter months. Fruit leather also makes a great gift, so it is worth making a large supply.

Other options include a fruit sorbet or homemade ice cream, a fruit purée providing the optimum base for both. My favorite for this is the cornelian cherry, which can be transformed into especially delicious creations. Fillings and the basic mixes for cookies, cakes, and muffins can also be prepared using fruit purée. Whole fruit can also be used, of course. Both can be used or even combined for a high-proof, ratafia-style liqueur. Fruit purée can be preserved by cooking it with the appropriate quantity of sugar or gelling sugar and sealing it in airtight jars (see Rose hip and pear purée, for example, on page 153).

Drying berries

Berries can also be preserved through drying, thereby retaining their vital ingredients over a long period of time. Cut larger berries into thin slices or bite-size pieces; smaller berries can be left whole. Traditionally, berries for drying would simply be spread out on a dish towel or rack and left in a warm, dry, and airy place for several days. This energy-saving method does need luck with the weather, though, ideally with temperatures of 62 °F (30 °C) or more. What is important is that you leave enough space between the berries so that the air is able to circulate. Turn the fruit over occasionally with a wooden spoon. Another method, one suitable for bad weather and for fruit with a high water content such as strawberries or raspberries, is to dry them in the oven. In this case, the berries are dried on parchment paper for 4–12 hours, at a maximum temperature of 83 °F (40 °C). Wedge the oven door open with a wooden spoon so that the moisture can escape. Slicing one of the berries will show you whether they are also dry on the inside. Professionals use a dehydrator or mechanical drier for drying.

Candied fruit

A further treat for the winter is candied fruit. This does require some effort, however. All kinds of berries and fruit are suitable for candying. Larger fruit varieties need to be sliced or cut into bite-sized pieces and the seeds removed, depending on the type of berry. Stone fruit is prepared by pricking the peel all over so that the sugar solution is able to soak in. For 2¼ lb (1 kg) fruit, combine 4 cups (1 liter) of water with 2¼ lb (1 kg) of sugar. Bring the sugar syrup to a boil and heat to over 212 °F (100 °C); 221 °F (105 °C) is ideal and can be reached by using a copper preserving pan, for example. The mixture should reach single thread consistency before adding the fruit and simmering for several minutes. Then remove the pan from the heat, cover, and leave the fruit to infuse for 24 hours. Remove the fruit pieces with a perforated spoon, reheat the syrup, and return the fruit to the syrup again. Repeat this procedure again (three times in total). The candied berries are then placed on a rack or on wax paper for about three days to dry. They should not harden, but should not be too sticky either. Store the candied fruit in an airtight container (e.g. in screw-top jars or sealed food containers) in a dark place. Placing wax paper between the individual layers prevents the fruit from sticking together.

Immunity through berries

Berries are known for their especially high vitamin content, particularly Vitamin C, but also vitamins A and E as well as B vitamins. They also contain health-promoting dietary fiber and many important minerals (including potassium, calcium, sodium, iron, and magnesium). They are therefore an essential component in a balanced diet.

Also of interest are the phytochemicals. These include the extensive group of polyphenols with flavonoids, anthocyanins, and carotenoids among others. These are visible coloring agents of considerable importance to our health. Flavonoids include flavones, found in yellow plants and fruit parts, and anthocyanins, which occur in abundance as red-blue coloring agents in blackberries, blueberries, elderberries, and black currants, for example. Other important coloring agents are the carotenoids, which are evident in yellow, orange, and red coloring. The most common carotenoid is beta-carotene, a preliminary stage of Vitamin A. These coloring agents are important for our health. Firstly, they are all antioxidant, meaning that they intercept harmful free radicals in the body—oxygen molecules that are chemically highly reactive and which can severely damage our health, especially at stressful times. Polyphenols have a stress-reducing effect and are anti-inflammatory in nature, protecting our cells and blood vessels and combating cardiovascular disease. A long-term study in the USA shows that ample consumption of blueberries or huckleberries stimulates blood vessel regeneration through the anthocyanins they contain, reduces high blood pressure, and significantly reduces the risk of heart attacks.

Polyphenols also detoxify, reduce blood pressure, relieve cramps, and improve the circulation. According to current research, they protect us from a variety of illnesses, including arteriosclerosis, heart attacks, and Alzheimer's disease. The immune system receives an all-round boost, while these active agents are even used in cancer prevention.

Tannins are also to be found in fruit and other plant parts. They are primarily astringent, antibiotic, anti-inflammatory, pain-relieving, and have a wound-healing effect. Tannins strengthen the blood vessels and therefore help to combat cardiovascular disease in particular.

Last, but not least, there are the essential oils that stimulate both body and soul through the olfactory senses. The olfactory center, part of the limbic system, is where feelings are based, instincts activated, and hormone release stimulated. Aromas play an important role in our well-being, an integral part of our delight in favorite treats and happy memories from childhood through to old age.

Growing berries yourself

The soil

Growing berries in your garden or planters can yield a seasonal variety of untreated, freshly harvested fruit for you and your loved ones. But you do need to observe a few rules in order to be able to harvest a well-filled basket year after year.
The gardener's most important element is earth. Berries are perennial and can flourish in the same place for many years, which means that the soil must meet all the requirements for optimum growth. Most berries need their root growth to be left as undisturbed as possible and therefore require sufficient space to flourish.

Balcony gardeners should pay particular attention to using structurally stable, organic potting soil, without peat. In a deep container of sufficient size, this will make a good home for berry plants for many years. Regular watering and moderate feeding with an organic liquid fertilizer are important. Different berries can also be combined in large containers: e.g. red currants with strawberries planted beneath them. This is a wonderful way to turn your balcony into a little paradise of berry delights.

The soil in the garden ought not to be too stony, but should not be too heavy either. Do the spade test to check your soil. Dig your spade into the ground with a hefty swing; if up to two-thirds of its blade can be inserted into the ground without much resistance, you have a soil well suited to good root development. Make a repeated cut of the spade in the planting hole to double-check the looseness of the soil.
You should now have a piece of soil shaped like a slice of cake. A dark color indicates active, growth-stimulating soil life. The organisms living here are the best garden aids, and should be cherished. To this end, simply mix the soil from the planting hole with some compost and you are ready for planting.
Compost is a complete fertilizer containing all the necessary nutrients, so no further fertilizing is necessary. Most berry bushes have shallow roots and are able to make full use of compost. For an entire season, 3 quarts (3 liters) of compost per square yard (per square meter) are adequate.

Once the berry plants have been in the same place for several years, you can give them an annual dose of berry fertilizer. For the long-term supply of nutrients, I prefer organic fertilizers. With these, I can be sure that no undesirables are being released into the soil that will later present in the berries as well. Organic, slow-release fertilizers have a soil-reviving and plant-boosting impact over the longer term. This ensures consistently balanced

growth, uniformly firm fruit shoots, and good winter hardiness.

Did it take a great deal of effort to get that spade to penetrate the soil in your garden? If so, it is likely to be a heavy, clay soil. Wonderful—clay soils are among the most fertile in the world. However, the soil structure will need to be improved and aerated in order to realize its potential. To do this, mix the clay soil from a generous planting hole with one part compost and a further part sand or lava granulate. You will then be ready to plant with confidence.
Special conditions apply to the substrate for blueberries, cranberries, and lingonberries. They need a soil with a sour (acidic) pH-value. To achieve this, mix the planting soil with coniferous wood shavings, coniferous bark mulch, and oak leaf mulch. Add a little organic sulfur (available in pharmacies, less than an ounce is sufficient). Cover the surface with a generous quantity of bark mulch. It is important that the soil moisture is checked regularly, as the berry bushes could otherwise suffer drought damage to the leaves and fruit.

Mulching and watering

A mulch layer of about 2 inches (5 cm) deep over the surface of the soil corresponding to the root radius is recommended, in order to promote soil life. Mown grass, hedge clippings, leaves, straw, bark or garden compost, or simply shredded garden waste, is ideal for this. Horn shavings should be applied as a fertilizer beforehand if using woody mulch.

Mulching has benefits for both plants and gardener: the surrounding weeds competing for water and light are largely eliminated, while hoeing and weeding are kept to a minimum. There is also less need for watering because the earth remains moist for longer—ideal conditions for the lazy bed gardener. Water is an important factor, particularly when the fruit is ripening. A lack of moisture during this time means that the fruit will remain small. It can even mean that the berries drop off. Entirely natural, however, is the "June fruit drop" when the berry bushes control the rejection of fruit themselves so as to continue developing only optimal berries.

The right location and natural aids

In order to be able to harvest as many sweet berries as possible, you need to choose a sunny, wind-protected location. Sun and warmth stimulate sugar production and promote good, even ripening. Wind-protected sites shield the branches from strong gales and allow the fruit to remain hanging without damage. Late frost, which can kill off the plant, derives from a combination of extremely cold temperatures and strong wind. Protected sites prevent this and even highly frost-resistant types like chokeberries or goji berries flourish better in such places. In short: the sunnier and more protected a location, the more successful the harvest should be.

In order to achieve good pollination and thereby a good fruit set with early blossoming varieties in particular, but also with other varieties, I have hung self-made, wild bee hotels all over my garden. These quickly attract mason bees, which, according to my observations, pollinate berry bushes much more effectively and earlier than honey bees. Professional growers have been utilizing this effect for more than ten years. Bee hotels are easy to build and provide a decorative feature for the garden.

Growth forms and pruning

Pruning is important if a berry bush is to stay healthy and produce plenty of delicious fruit for as many years as possible. The recommendations differ according to plant variety and gardener. I personally check my bushes directly after the harvest for obsolete shoots and cut off any harvested shoots more than two years old—recognizable by their darker color. This thinning out means more sunlight reaches the plant so that, the following season, there is increased bud formation and therefore a richer harvest. Plant disease can also be prevented through pruning. Red currants, gooseberries, blackberries, and raspberries in particular require regular rejuvenation.

A special form of bush cultivation is trellis or espalier growth, where just a few shoots are fastened individually to a frame and trained along it. All of the other branches are cut back. The remaining fruit shoots enjoy optimum development, because they are able to ripen in the sun and set fruit on all sides. These often hang over the foliage, making harvesting much easier.
Large bushes or small trees from wild hedges (e.g. barberries or medlars) should be checked regularly for dead wood and shade-forming branches. Regenerative pruning can be carried out just as well in winter with these varieties.
It is important to feed a bush with organic berry fertilizer after pruning. This enables the plant to recover, setting new shoots and, more especially, fruit buds.

A particularly attractive and also practical form of berry bush is the tall-stemmed bush. Red currants and gooseberries in particular are readily available in this form, as are chokeberries and a variety of roses (rose hips). Planting tall-stemmed plants and bushes alternately and slightly set back from one another produces a very attractive overall effect. The differing heights can also be used to increase yields in smaller spaces.
Tall-stemmed berry bushes are generally very robust. In colder regions, however, it is worth applying white trunk paint at the beginning of the winter to prevent possible frost cracks in the trunk. The closely spaced shoots at the top end can also be wrapped in a covering of permeable white fleece. Berry bushes, on the other hand, are so frost-hardy that they do not require any winter protection.

Plant protection

Fortunately, with robust berry plants, plant protection is not a major issue. Fungal diseases like mildew tend to occur seldom when the location has been well chosen. In particular, the location should be airy, and afford sufficient space on all sides. In the event of an infestation, there are effective organic treatments available, while aphids and gooseberry sawfly can be held at bay with organic insecticides that break down quickly without a residue. Resourceful gardeners also plant wild garlic or few-flowered leek beneath their bushes, which have a vitalizing effect on the plants as well as reducing mildew and aphids.

Whether wild or garden berries, tall-stemmed or espalier, in beds or in containers, when purchasing berry plants it is usually worth seeking out a good, local, specialist nursery. This is where you will find professional advice and a broad spectrum of varieties encompassing different characteristics, ripening times, and location requirements—and the right berries for your own garden. I personally would only order by post in the case of container plants in winter. Opening plant packages does sometimes bring disappointment: as a result of the journey it could be softened shoots, damaged young foliage, or broken fruit buds that reach your door. The best time for planting berry bushes is fall or spring, and this is when most plants are also available locally.

Berries from the garden

Strawberries—everybody's favorite

We all love them—the fat, red, sweet fruit from the bed in the garden. They are to be found in every larger home garden and also on the smallest of balconies, in pots, boxes, and hanging baskets. This special fruit has been flourishing in human settlements since the Stone Age. For the Germanic peoples it was seen as a symbol for sensual pleasures and was dedicated to the fertility goddess Freyja. "Going down to the strawberry fields together" used to be a pleasantly worded invitation to intimacy. Hence Saint Hildegard of Bingen warned against the "depraved" berries—while folklore extolled their virtues.

Strawberries (*Fragaria*) are widely available and increasingly so. The first strawberries are already there to delight as of May, while the multiple harvest varieties produce our favorite fruit well into the fall. One forebear of the countless different varieties available today is the fruit of the wild strawberry (*Fragaria vesca*). The botanical name *Fragaria* means "fragrance" and *vesca* means "edible"—the edible fragrance. Together with the musk strawberry (*Fragaria moschata*) they were cultivated until the Middle Ages. It was only in the 18th century that explorers in North America discovered the Virginia strawberry (*Fragaria virginiana*), its large fruit ensuring that it was promptly brought over to Europe. The South American continent then supplied an aromatic newcomer, the Chilean strawberry (*Fragaria chiloensis*). Our garden strawberry (*Fragaria* x *ananassa*) is a hybrid of these two varieties.

Strawberries can be used for countless diverse and delicious culinary delights. They are at their very best freshly picked—and with a generous helping of cream, of course. In the summer, they are a treat in the form of ice cream or sorbet; and they help us get through the winter as jams, juices, and frozen fruit. Preserved in rum and sugar as the German dessert *Rumtopf*, they are available as a spirited treat all year round. No fruit salad is complete without them, while a strawberry gâteau is one of the most popular cakes there is. The German red fruit jelly *Rote Grütze* can be called such only when it contains strawberries, and any leaf salad can be enlivened with the addition of ripe strawberries, a little balsamic vinegar, and green peppercorns. Strawberries are also particularly welcome in a refreshing summer punch or drink, or in combination with a range of milk products such as quark, cream, mascarpone, and ricotta. You need no more than a few strawberries in the fridge to be able to conjure up something delicious!

The inner values are also significant: ripe strawberries contain more Vitamin C than lemons. A 7-ounce serving (about 200 g) is enough to meet our daily requirement of this vitamin. Other important elements include folic acid, the B vitamins, Vitamin K, calcium, potassium, magnesium, iron, zinc, manganese, and phosphorus. Also of relevance is the proportion of antioxidant substances (anthocyanin) with their range of preventative properties. Strawberries also have few calories, making them a guilt-free snack. The fruit have a diuretic, blood-cleansing, and invigorating effect. They boost the metabolism and are also said to lower blood pressure. The methyl salicylate they contain also means that strawberries even have a mild pain-relieving effect, capable of countering headaches or even migraines in some cases!

In the garden, strawberries need full sun with fertile soil and a regular water supply. The soil ought to be mulched to retain the moisture and to prevent the fruit coming into direct contact with the soil. Straw is ideal for this, but its color can delay the ripening of the fruit somewhat. A dark soil, with a top layer of *terra preta* (black earth), for example, allows the fruit to ripen much more quickly. With good care and healthy soil, you can leave your strawberry plants in the same place for several years. This also enables you to obtain runners from rare varieties or from your own favorite types. My favorites are the rare White Pineapple (pink-white fruit with a pineapple aroma) and the perpetual fruiting Mara des Bois with its flavor of wild strawberries all season long. It is diversity that provides us with this abundance—the greater the range of varieties flourishing in a bed, the better the fruit set and, ultimately, the harvest.

Wild strawberries are a special treat. They have a wonderfully intense aroma and, because they are one of the original, indigenous fruit varieties, they are especially robust. The fruit is much smaller, but they are unbeatable as a snack. I harvest these elongated strawberries carefully for punch, fruit juices, and especially for smoothies. The flesh is looser, making these berries pressure sensitive, so immediate use is best. This happens automatically for anyone with children in the garden.

My favorite varieties are the White Solemacher and its red sister variety, both of which are known worldwide. According to my taste, these are the best varieties for a light summer strawberry punch. They are the perfect accompaniment to vanilla ice cream, and made into a sorbet they will take you straight to culinary heaven.

Wild strawberries flourish in sunny spots with a humus soil. The occasional application of organic berry fertilizer will produce slightly larger fruit. It is advisable to plant the commercially available varieties guaranteed to carry fruit. Taken from an outdoor location, you can mistakenly end up with a dioecious musk strawberry with no fruit set.

Those wanting the full aroma of the wild strawberry together with the size of the garden strawberry in one fruit are best served by the rampant Florika strawberry. It can take over large areas of the garden all by itself.

Flavorsome cooking with strawberries

Strawberry and elderflower jam

MAKES 4 × 1 LB (454 G) JARS

2¼ lb (1 kg) strawberries
juice of ½ lemon
3 handfuls of elderflower, heads only
2¼ cups (500 g) gelling (jam) sugar

■ Wash the strawberries, remove their green tops, then chop the fruit and crush it lightly.

■ Place in a non-metallic bowl and stir in the lemon juice, elderflowers, and gelling sugar. Leave to infuse for at least 3 hours. Transfer to a preserving pan and bring to a boil, stirring all the time. Simmer for at least 4 minutes, still stirring, then test for setting.

■ Pour into clean, sterilized jars, filling them right to the top, and seal immediately. Turn the jars upside down for several minutes, then turn them back up and leave to cool.

Summery strawberry-and-coconut gâteau

MAKES 1 GÂTEAU (9½-IN/24-CM DIAMETER)

For the base
6 oz (175 g) white couverture chocolate
2 tbsp cream
3½ oz (100 g) cornflakes
½ cup (50 g) shredded coconut

For the topping
1¾ lb (750 g) strawberries
6 gelatin leaves
1 cup (240 ml) coconut milk
⅔ cup (75 g) confectioners' sugar
juice and zest of 1 unwaxed lime
1 cup (100 g) shredded coconut
2 cups (500 g) low-fat quark
generous ¾ cup (200 ml) whipping cream
4 tsp (20 ml) coconut liqueur

For the garnish
shredded coconut
mint leaves
confectioners' sugar

■ For the base: line a springform pan with parchment paper. Roughly chop the couverture chocolate and melt it together with the cream in a bowl over hot water, or in a bain-marie.

■ Place the cornflakes in a freezer bag and crush with a rolling pin. Remove the chocolate mixture from the heat, and combine with the cornflakes and shredded coconut.

■ Spread evenly over the base of the springform pan, and leave in the refrigerator for about 30 minutes to set.

■ For the topping: wash the strawberries and remove the green tops. Cut the strawberries in half, then layer half of them over the base of the pan. Set 10–12 strawberries aside for the garnish, then purée the rest and pass through a fine sieve.

■ Soften the gelatin in cold water. Mix the coconut milk with the confectioners' sugar, lime juice and zest, the shredded coconut, and the quark, and stir until smooth. Whip the cream until stiff.

■ Place the softened gelatin in a small pan, together with the liqueur, and dissolve over low heat. Stir in 2–3 tablespoons of the coconut mixture, then fold in the remainder before gently folding in the whipped cream.

■ Spoon the mixture into the springform pan, smooth the surface, and arrange the remaining strawberries on top. Chill in the refrigerator for at least 3 hours.

■ Remove from the pan before serving, drizzle with the strawberry purée, and garnish with shredded coconut and mint leaves. Finally, dust with confectioners' sugar.

Blackberries—sweet, black bundles

A thicket full of large, thorny bushes, covered in shiny, black berries—such was the state of the piece of land when we bought it, where our house and nursery now stand. I can still taste the wonderful flavor of the bumper harvest to this day. Left to grow wild, the blackberry (*Rubus fruticosus*) will quickly establish impenetrable hedges along pathways, railroad tracks, forests, and around abandoned properties. In the past, instead of erecting a fence or a wall to protect their settlements, people planted blackberry bushes around their villages. Also known as brambles, these prickly plants afford protection, tea, and vitamins on the doorstep. For today's gardener, though, it is advisable to keep the bushes in check.

Wild blackberries are full of flavor and so it was only in around 1850 that people came up with the idea of cultivating new varieties, the intention being to combine the fine flavor with a thorn-free shrub. It was not until around 1930, however, that the first, really successful ones of this type came onto the market, with the well-known "Thornless Evergreen" still available today.

The sweet, black bundles ripen between July and October. Harvested too early, they lack their full flavor and are acidic. Once almost overripe, though, they taste fantastic and sugary sweet. By this time, they will also be soft and pressure sensitive, so should be used as quickly as possible. Harvesting, processing, and tasting all leave their mark on fingers and tongues. The blue coloring

agents remain on the skin for quite a while, and stains on clothing are difficult to remove. Wearing protective clothing against the thorns is also advisable when harvesting.

There are numerous possibilities for using blackberries: in addition to cakes, gâteaux, jams, and jellies, they are ideal for chutneys, fruit sorbets, syrups, smoothies, cold desserts, tarts, ice

creams, and milkshakes. I am especially fond of them baked in a clafoutis—a French dessert tart—or cobbler. They can also be combined with distilled spirits (e.g. vodka or gin) to make fine liqueurs such as "Crème de Mûre," which makes a wonderful basis for cocktails.

Fruit sauces made from blackberries go well with desserts and savory meat dishes.

A fruit purée is another good way of processing blackberries, which are first cooked briefly to soften them. Sugar can be added while they are cooking, depending on what you plan to use them for. They are then passed through a sieve or food mill. This removes the fibrous central cores, known as "receptacles;" unlike raspberries, the receptacles remain in blackberries when they are picked. The resulting fruit purée can also be used to make all of the culinary delights listed above. It also freezes well, as do the fresh berries.

Blackberries are rich in tannins, vitamins A, C, and E, and also minerals such as calcium, potassium, iron, and magnesium. As free radical neutralizers, coloring agents like flavonoids and beta-carotene boost the immune system. The connective tissues and vascular walls, in particular, respond positively to the active agents in blackberries. The fruit also supplies dietary fiber, which is important for a healthy digestive system. The medicinal effectiveness of a flavorsome tea made from blackberry leaves in combatting diarrhea is scientifically proven. Drinking and gargling with blackberry tea or juice are recommended for mouth ulcers, sore throats, and hoarseness.

In the garden, blackberries flourish in sunny through semi-shaded, frost-protected locations, sheltered from the wind. This climbing plant bears its fruit on two-year-old shoots. When pruning, you therefore need to leave enough young shoots in place for next year's harvest. A wire frame will make future harvesting easier: fasten a total of 3–5 young shoots per plant in even arches. This will make even the thorny varieties less threatening. All of the shoots close to the ground should be cut off following the harvest, to ensure that the interior of the bush receives enough sunlight. The blossoms will be all the more abundant the following year.

Apple and blackberry pie

SERVES 4–6

For the shortcrust pastry

1/3 cup (50 g) all-purpose flour
1/3 cup (50 g) spelt flour
5 tbsp (70 g) butter
3 tbsp (40 g) sugar
1 egg yolk
1–2 tbsp cream

For the filling

4 apples
7 oz (200 g) blackberries
1¼ cups (300 ml) apple juice
5 tbsp (40 g) cornstarch
3½ tbsp (50 g) sugar
2 tbsp vanilla sugar
1 egg white
confectioners' sugar

■ Place all of the ingredients for the shortcrust pastry in a bowl and work until smooth, using the dough hook of a food processor. Cover the dough in plastic wrap and chill for about 30 minutes.

■ Pre-heat the oven to 390 °F (200 °C). Peel the apples, quarter them, remove the cores, and dice. Pick any stalks or leaves from the blackberries, wash the fruit, and pat it dry with paper towels.

■ Place about ¾ cup (200 ml) of the apple juice in a saucepan and bring to a boil. Combine the remaining juice with the cornstarch, sugar, and vanilla sugar, and stir until smooth. Pour into the boiling apple juice, stirring all the time, and bring back briefly to a boil. Stir in the apple pieces and the blackberries, remove from the heat, and leave to cool slightly.

■ Beat the egg white until stiff and fold into the fruit mixture. Pour the mixture into a deep, ovenproof dish or baking pan (around 10 inches / 26 cm in diameter).

■ Roll the pastry into a circle the size of the baking pan and lay it loosely over the filling. Press the edges onto the edges of the pan. Sprinkle with sugar and bake in the pre-heated oven for about 40 minutes.

Flavorsome cooking with blackberries

Blackberry and lime mousse

SERVES 4

5 red gelatin leaves
14 oz (400 g) fresh blackberries
3 tbsp water
juice and zest of 1 unwaxed lime
7 tbsp (100 g) sugar
10 tbsp (150 ml) whipping cream
2 egg whites
1 unwaxed lime, sliced

■ Soften the gelatin in cold water. Pick any stalks or leaves from the blackberries and wash the fruit gently. Keep a few, nice berries aside for the garnish.

■ Place the rest of the blackberries in a saucepan with the water, lime juice, and sugar. Bring to a boil and simmer over low heat for about 8 minutes, then pass through a sieve or food mill.

■ Squeeze the water from the gelatin leaves then add them to the blackberries, stirring until dissolved. Chill the mixture in the icebox for about 90 minutes until it starts to set, stirring occasionally; do not allow it to become too firm.

■ Whip the cream together with the lime zest, until stiff. Beat the egg white until stiff. Fold both into the berry mixture. Line individual molds with plastic wrap, fill with the mousse, and chill for about 3 hours.

■ To serve, carefully turn the mousses out of the molds onto dessert plates, and remove the plastic wrap. Garnish each mousse with lime slices and the reserved blackberries.

Blackberry cobbler

SERVES 4

$^2/_3$ cup (100 g) all-purpose flour
½ cup (50 g) ground almonds
1 tsp baking powder
$^1/_3$ cup (75 g) sugar
pinch of salt
3½ tbsp (50 g) butter
¼ cup (60 ml) cream
generous 1 lb (500 g) blackberries
2 tbsp cornstarch
softened butter, for greasing
2 tbsp vanilla sugar

■ Combine the flour with the almonds, baking powder, 3½ tablespoons (50 g) of the sugar, and the salt in a bowl. Cut the butter into pieces and rub it into the flour mixture until it resembles breadcrumbs.

■ Then quickly fold in the cream until the dough becomes light and sticky. Pre-heat the oven to 350 °F (180 °C) (fan).

■ Pick any stalks or leaves from the blackberries, rinse the fruit carefully, and pat dry. Combine the berries with the cornstarch and place them in a greased, ovenproof dish or in individual baking pans. Sprinkle with the remaining sugar and vanilla sugar combined.

■ Tear the dough into pieces and spread over the blackberries. Bake in the oven for about 30 minutes until golden brown. Goes well with vanilla ice cream.

Currants—diversity in red, white, and green

Around about the time of the Feast of St John (June 24), the keen gardener is likely to be strolling through his or her little piece of paradise, checking to see whether the currants (genus *Ribes*) are ripe. This popular sweet-and-sour fruit is recommended for novice growers as the plants are undemanding, low-maintenance, and robust, while pruning is also easy. They even grow well in containers on balconies and terraces, the tall-stemmed bushes being especially attractive.

It was via the monastery gardens of the 14th century that currants became widespread. Their popular name in Germany, the gout berry, refers back to an ancient healing ritual in which gout was symbolically bound to the bush with a blessing and colored ribbon. The plant was believed to protect against further attacks, as well as being a cure. The monks later linked this custom with Christian rituals. They crossed the wild, indigenous white currants (*Ribes petraeum*) with red and white garden currants (*Ribes rubrum* var. *domesticum*)—on account of their color, berry size, and higher yield. And so, over the centuries, the sweetest and the largest berries were selected. Currants have been grown on a large scale for commercial use since 1650; for example, in the production of fruit wine.

In addition to currant wine, these vitamin-rich berries can also be used to make juice, cordial, liqueur, syrups, and shrubs. A shrub is made from fruit, vinegar, and sugar—and makes a wonderfully

refreshing summer drink. For me, no red fruit jelly can be without the fine, sweet-and-sour aroma of currants. If you find the red berries too sour, try the white or green varieties (e.g. *Ribes grossularia*, or gooseberry): they are milder and sweeter. Instead of anthocyanin, they contain flavonoids, which color the dish yellow instead of red.

Currants can be kept in the refrigerator for a while after harvesting or purchase. It is best to place them loosely in a bowl, as the berries are somewhat pressure sensitive. With their high pectin content, they can be used to make jams and jellies without the use of artificial gelling aids. Just use granulated sugar (at a ratio of 1:1) and allow the mixture to boil for longer. Currants are also a useful gelling agent in recipes using other fruit combinations.

The plants are self-pollinating; several bushes should therefore result in abundant pollination, more berries, and, in particular, a longer harvest period. A good water supply is important once the fruit starts to ripen, otherwise the berries will drop, unripe, from the bush. Provided you pay attention to this, you will be able to harvest from June through late summer. The berry clusters, with their differing lengths, need to be picked carefully or the berries will drop off according to their degree of ripeness. A pair of small pruning shears is ideal for this, so that you can cut the berry clusters off cleanly at the stem.

Fourberry—a new addition

You have probably never heard of the fourberry: this new addition is a very special variety of the golden currant (*Ribes aureum*). The latter is often planted in parks and gardens on account of its blossom, while the fourberry bred from it is also impressive in the spring, with its wonderful, intensely yellow clusters of blossoms. Its strong fragrance envelops every passer-by in a cloud of scent. From mid-July, the bush bears marble-size, shiny, black berries with their very own flavor.

It was a Swiss grower who came up with the idea of creating this fruity delight from a wild specimen. Over the course of ten years, he selected one thousand seedlings and isolated the best, the largest, and the tastiest. The result first came to market in 2011: to look at, the fourberry resembles the jostaberry, but, unlike the latter, does not taste like a gooseberry. Its aroma is much fruitier, juicier, and with a very interesting, characteristic, resinous nuance. These berries also need to be fully ripe to have their full, sweet taste. They practically need to fall into your hands by themselves.

The grower is working on further innovations. The next few years will see another fourberry with a very unusual fruit color: bright orange. Only a few fruit, such as the Cape gooseberry (*Physalis peruviana*), have a similar color, so this will open up brand new perspectives in the kitchen. We will also be able to use these varieties to make wonderfully orange-color jams, juices, and other such bright delights. I am already looking forward to enjoying these fourberries, perhaps in combination with red currants, black currants, and green gooseberries—ideally, all together in a fruit salad. So much color just has to be good for you!

Growing up to 5 feet (150 cm) in height, the fourberry bush is an eye-catcher in both fruit and flower gardens. The abundant, fragrant blossoms in spring are followed by the tasty, black berries, while in the fall the bushes come up trumps again with orange through dark-red leaves. Golden currants originate from the North American prairie, which means that fourberries are robust and easy to care for. The bush works well both as an individual specimen or gap filler. A loose garden soil, not too wet, and regular doses of an organic berry fertilizer ensure even growth and healthy new shoots. Older shoots, i.e. from the age of four years, should be cut out. This thins out the bush, which in turn promotes bud formation and fruit set.

Golden currants are a tried and tested rootstock for grafting currant bushes. They blossom from the first year onward, so you can expect an early first harvest. The fruit ripen consecutively, so the home grower can harvest fresh berries daily for several weeks for snacks, salads, or as a topping for muesli. Fourberries also flourish in containers on balconies and terraces. In this case, it is advisable to position the plants against a house wall in the winter, because a heavy frost could damage them.

And, should you happen to miss the harvest while away on holiday, the berries dried on the bush are ideal for use as raisins.

Black currants—cassis and more

Opinions differ when it comes to black currants (*Ribes nigrum*). Some people find even the fragrance unpleasant, while others enthuse about the unique aroma of the "cassis berry." Its wonderful health benefits are undisputed, however. In some cultures, it is even symbolic of long life.

In central Europe the plant, characterized by vigorous growth in the wild, flourishes especially in damp, alluvial forests. Its characteristic smell, evident just by rubbing the leaves and bark, distinguishes the black currant from any red currant (*Ribes rubrum*) or white currant (*Ribes petraeum*) bushes growing in the same location.

In the past, the plant was commonly used for medicinal purposes, taken as a tea to aid digestion and counter constipation. Gardeners selected the tastiest specimens and propagated them. This is how, over time, these shiny, blue-black berries found their way into the kitchen and home garden, where they have been an established feature since the 16th century.

In earlier times of self-sufficiency, the gathered berries were often used to make a liqueur, combining them with sugar and distilled spirit in a bottle. The result was crème de cassis: the famous, dark, finely aromatic liqueur. One of my surprising taste experiences was a red currant ice cream served with crème de cassis as a sauce—delicious! It can be used to refine a great many desserts, soups, jams, and cakes. Juices can also be enhanced with its characteristic aroma. Combine crème de cassis with Champagne, and you will be serving Kir Royale.

Low in calories, black currants contain three or four times as much Vitamin C as lemons and are also far superior to red currants in this regard. They have a high terpene content, as well as anthocyanin, flavonoids, B vitamins, minerals, pectin, tannins, and citric acid. According to latest research, kernel oil from black currants is rich in gamma linoleic acid. This fatty acid is used in particular to boost the immune system in respect of both the skin and the body. As noted above, dried black currant leaves can be infused like a tea: they are antioxidant, diuretic, and have a mild blood pressure reducing effect. They are also ideal as an ingredient for wild herb soup.

In the garden, black currants are ideally suited to dark, slightly damp locations. The selective pruning of aging shoots keeps the plant healthy and stimulates growth. The particular frost hardiness of these indigenous berries is evidenced by the fact that they are grown in Finland and Poland. The plant is also suited to growing in pots on balconies and terraces.

My recommended varieties: page 179.

Crème de cassis

Makes 4 × 8-oz (250-ml) jars

2¼ lb (1 kg) black currants
1¼ quarts (1.2 liters) dry red wine
2¼ lb (1 kg) gelling (jam) sugar (1:1)

■ Carefully rinse the black currants, remove from their stalks, and leave to drain well. Place in a saucepan and pour in the red wine.

■ Lightly crush the black currants with a potato masher, and leave to infuse for about 2 hours.

■ Stir in the sugar, bring to a boil, and simmer for about 5 minutes, stirring all the time.

■ Pour into clean jars and seal well. Turn the jars upside down for a few minutes, and leave to cool the right way up.

Flavorsome cooking with currants

Red currant juice

Makes 2 cups (500 ml)

generous 1lb (500 g) red currants
generous 1lb (500 g) sugar
2 cups (500 ml) water
1 tbsp lemon juice

■ Pick through the red currants, wash them gently, and pat dry with paper towels. Remove the stalks.

■ Combine the sugar with the water and lemon juice and bring to a boil. Simmer for about 5 minutes, stirring occasionally. Remove from the heat, add the berries immediately, and crush lightly with a potato masher.

■ Cover the pan and leave to infuse for at least 4 hours. Pass through a fine sieve or food mill, retaining the liquid. Bring back to a boil and pour into clean, sterilized bottles.

■ Seal well, and leave to cool.

Currant sorbet

Serves 4

1 vanilla bean
1¼ lb (600 g) red currants
6 tbsp sugar
¾ cup (200 ml) ice-cold currant or grape juice
confectioners' sugar, for dusting

■ Slit the vanilla bean open lengthwise, and scrape out the seeds. Pick through the red currants, wash them, and pat them dry.

■ Set aside four clusters for the garnish, then remove the rest of the currants from their stalks. Freeze for about 2–3 hours.

■ Purée the frozen fruit with the sugar, vanilla seeds, and ice-cold juice. Place the sorbet immediately in chilled glasses, garnish with the reserved berry clusters, and dust with confectioners' sugar.

Variation

If you are not catering for children, you could replace some of the juice with the same amount of crème de cassis.

Black currant jelly

Makes 1 loaf pan (4 × 8 in / 10 × 20 cm)

1¾ lb (750 g) black currants
1 vanilla bean
2 heaped tbsp agar-agar
¾ cup (200 ml) black currant juice
scant 1½ cups (300 g) sugar
1 tbsp lemon juice

■ Pick any leaves and stalks from the black currants, wash the fruit, and leave to drain well.

■ Slit the vanilla bean open lengthwise, and scrape out the seeds. Combine the agar-agar with 2 tablespoons of currant juice.

■ Lightly crush the currants in a saucepan with a potato masher.

■ Combine with the rest of the currant juice, the agar-agar, and the sugar, and bring to a boil. Simmer gently for about 2 minutes, stirring frequently.

■ Remove the saucepan from the heat, then add the vanilla seeds and lemon juice.

■ Line a loaf pan with plastic wrap, and fill with the currant mixture.

■ Leave to set in the refrigerator for about 4 hours. Then tip out onto an elongated serving platter and serve with bread and cheese.

Turkey roulade with red currants

SERVES 4

4 turkey scallops (about 6 oz / 150 g each)
salt and freshly ground black pepper
7 oz (180 g) quark
1 tbsp lemon juice
4 oz (100 g) red currants
2 tbsp chopped thyme leaves
2 tbsp rapeseed oil
7 tbsp (100 ml) dry white wine
1 cup (250 ml) chicken stock
2 tbsp currant juice
thyme sprigs, for garnishing

■ Wash the turkey scallops, pat them dry, flatten with a meat pounder, and season with salt and pepper.

■ Mix the quark with the lemon juice and season lightly with salt and pepper. Remove any leaves or stalks from the red currants and wash the fruit. Set aside about 1 tablespoon of currants for the garnish.

■ Fold the rest of the currants into the quark mixture, together with the chopped thyme leaves. Spread over the turkey scallops, roll them up, and fasten with kitchen twine.

■ Heat the oil in a skillet and brown the roulades on all sides until golden brown. Deglaze with the wine and stock.

■ Add the currant juice to the skillet, cover, and leave to simmer over low heat for about 30 minutes. Remove the roulades and cut off the twine. Reduce the sauce in the skillet, uncovered, and season with salt and pepper.

■ Arrange the roulades on warmed plates, with some of the sauce. Garnish with thyme and a few fresh currants.

Currant dressing

SERVES 4

1 handful each, black currants and red currants
2 tbsp honey
2 tbsp balsamic vinegar
1 tbsp Dijon-style mustard
8 tbsp olive oil
salt, pepper

■ Pick any leaves or stalks from the currants, wash the fruit, and pat dry.

■ Place in a blender, and purée together with the honey and balsamic vinegar. Then pass through a sieve if you prefer to remove the seeds.

■ Beat the currant purée together with the mustard and olive oil, and season with salt and pepper.

Gooseberries—furry outside, sweet inside

Sour? Nothing of the sort! The European gooseberry (*Ribes uva-crispa*) has the highest sugar content of all berries after grapes. Provided the berries are fully ripe, of course.

These berry bushes are a widespread feature of the natural environment—you can find them growing or running wild, their preferred location being along rocks and canyons. In hedges too, on alluvial soil, and in deciduous forests they are a conspicuous sight in the spring, with their new, bright green foliage shining between the trees. The development from thorny wild fruit with arching shoots to the compact, low-maintenance berry delight we know today was swift. There were already 500 varieties available in German collections in 1850. Commercial growing began in 1896 with 26 varieties in the Lower Rhine region, with every region subsequently developing their own local specialties.

In the 1930s, however, gooseberries throughout Europe fell victim to the fungal disease known as American gooseberry mildew. This was introduced via plants from the USA, and indigenous varieties were helpless against it. American gooseberries, though, were resistant, so all of the newer European cultivars carry these genes.

For me, the fully ripe, sugary-sweet berries with their slightly sour edge were always the highlight in my grandma's meringue gâteau. There are not only traditional recipes to be found in my

kitchen, however. You will also come across a gooseberry jelly with lavender, and sweet-and-sour chutney with zucchini, chili peppers, or ginger. Gooseberries combine well with strawberries, kiwis, and bananas—in smoothies, for example, or jam.

In the 1930s, gooseberry wine was the pride of every community gardener, the natural sweetness being used to produce a strong, aromatic fruit wine. The bushes are often overburdened with

berries and it was soon discovered that unripe berries can be harvested too. These "green pickings" lighten the bushes' loads and allow the remaining fruit to ripen better. Preserving them with sugar produces a fine, durable base for all kinds of cakes and desserts. The green berries remain firm and yet still develop their wonderful aroma. In addition, gooseberries contain a great deal of Vitamin C, folic acid, and silicic acid.

There is always a spot to be found in the garden for an attractive gooseberry shrub or tall-stemmed bush. The plant also grows well in containers. Gooseberries do not require full sun and will also flourish in front of other bushes or in darker corners. The tremendous sweetness of which they are capable, though, develops best in a sunny location. In terms of soil, gooseberries are very low-maintenance as they can cope with just about every garden soil. Regular watering ensures better fruit size. What is important is that the older, woody shoots are cut away. Thinning out is best done directly after harvesting, and should leave only the full bearing, younger shoots in place. Mildew attacks are possible in very hot years—organic antifungal agents or a mixture of baking powder, washing-up liquid, and water will help against that. Sprayed on the leaves, it will naturally halt the advance of this white fungus.

Flavorsome cooking with gooseberries

Red gooseberry jam

Makes 4 × 8-oz (250-ml) jars

1¾ lb (800 g) red gooseberries
¾ cup (200 ml) red grape juice
2¼ lb (1 kg) gelling (jam) sugar (1:1)

■ Wash the gooseberries, drain, and place in a large saucepan together with the grape juice and gelling sugar.

■ Lightly crush the gooseberries with a potato masher and bring to a boil. Cook over high heat for about 5 minutes, stirring all the time. Test to see whether setting point has been reached by placing a few drops of the jam on a cold plate. If the drops become firm, the jam is ready.

■ Pour into clean, sterilized jars. Seal well and turn the jars upside down for a few minutes. Then turn them the right way up again and leave to cool.

Baked trout with vegetables and gooseberries

Serves 4

4 oven ready trout, 11 oz (300 g) each
salt and freshly ground black pepper
2 zucchini
2 ripe beefsteak tomatoes
11 oz (300 g) gooseberries
4 tbsp olive oil
2 tbsp fresh oregano, chopped

■ Pre-heat the oven to 390 °F (200 °C). Wash the trout and pat dry. Season with salt and pepper, both inside and out.

■ Wash the zucchini, cut in half lengthwise, and slice thinly. Wash the tomatoes, cut in half, remove the core, and then slice thinly. Pick any leaves or stems from the gooseberries, wash the fruit, and leave to drain.

■ Brush 4 pieces of aluminum foil (about 12 x 12 in / 30 x 30 cm) with 1 tablespoon of olive oil each. Place a trout on each and arrange the zucchini, tomato, and gooseberries on top. Season again with a little salt and pepper, fold the aluminum foil into a parcel, and seal well.

■ Place the 4 foil parcels on a baking sheet and cook in the pre-heated oven for about 30 minutes. Remove from the oven, sprinkle with the chopped oregano, and serve.

Jostaberries—juicy and sour

The jostaberry (*Ribes nidigrolaria*) is a cross between black currants and gooseberries. The aim of the hybridization was to combine the wonderful, cassis flavor of the black currant with the fruit size of the gooseberry. The juicy, red-black fruit have the properties of both parents and the fruit size is precisely midway between the two. The arrangement of the berries in clusters comes from the black currant plant, as does the lack of thorns, the remarkably high Vitamin C content, and the important anthocyanin. Jostaberries combine the fine, sweet-sour aroma of gooseberries with the slightly bitter flavor of black currants.

The imposing bushes bear a great many relatively large berries in July / August, supplying a sure basis for mixed jams and jellies, liqueurs, juice, red berry jellies, or a German "Rumtopf" (mixed fruits preserved in sugar and rum). Their high pectin content facilitates setting. Combined with a sweet apple juice and chilled mineral water, jostaberry syrup or juice makes for wonderful summer refreshment. I also like to make jostaberries into little cakes or cookies. In a fruit salad, or mixed with yogurt, quark, and sugar, they produce an aromatic dessert. Vanilla combines well with the slightly bitter berries. Their flavor is not for everyone, but for me just enjoying them fully ripe direct from the bush is a delight—for the blackbirds, too, so I have to be quick! Topped off with a goodly portion of whipped cream with vanilla sugar, the pleasure is complete.

Jostaberries are ideally suited to freezing, as they retain their quality and shape. This also means that they can be enjoyed later in the year, when the berry glut is a distant memory.

Even when fully ripe, jostaberries remain relatively firmly on the bush. You need to pick them carefully, so as not to squash the fruit in your hand. This is the reason they cannot be picked by machine. As such, they are rarely seen in stores, but are often available at farmers' markets. Growing your own bushes, which can grow to nearly 7 ft (2 m) in height, is certainly worth doing. Jostaberries require little maintenance, and yet they produce high yields. They have the advantage of a strong resistance to disease and pests. The bushes are as undemanding as their parent plants, but have also inherited their watering requirements when the fruit is ripening. Mulching is a good way of keeping water in the ground for longer, covering the shallow roots and ensuring the supply of nutritional humus.

Their vigorous growth does mean that pruning to thin them out becomes necessary from the third year onward. Do this in winter by cutting off the old, already harvested shoots, directly above the ground; otherwise, they will branch out too much and create a jungle, so that harvesting becomes difficult. Six to eight main shoots are enough to secure a good crop the following year; it is difficult to stop a jostaberry from flourishing, even if the occasional pruning error is made.

Apple jostaberry muffins

Makes 12 muffins

3½ oz (100 g) jostaberries
flour, for dusting
1 large apple

For the muffin batter
1¾ cups (250 g) all-purpose flour
2 tsp baking powder
½ tsp bicarbonate of soda
1 tsp cinnamon
1 egg
7 tbsp (100 g) sugar
⅓ cup (80 ml) vegetable oil
1 tbsp vanilla sugar
1 cup (250 ml) buttermilk

For the crumble topping
7 tbsp (100 g) butter
⅔ cup (100 g) all-purpose flour
generous ½ cup (50 g) ground almonds
7 tbsp (100 g) sugar

confectioners' sugar, for dusting

■ Line the muffin pan with paper cases. Pre-heat the oven to 350 °F (180 °C).

■ Pick through the jostaberries, then wash them and pat dry. Dust with a little flour. Peel the apple, cut it into quarters, remove the cores, and grate finely.

■ For the muffin batter, combine the grated apple with the flour, baking powder, bicarbonate of soda, cinnamon, and jostaberries in a bowl.

■ In a separate bowl, whisk the egg and mix together well with the sugar, oil, vanilla sugar, and buttermilk. Fold lightly into the flour and apple mixture until the liquid is evenly distributed.

■ For the crumble topping, combine all the ingredients and work into crumbs with your hands.

■ Divide the muffin batter between the paper cases and sprinkle the crumble topping on top. Bake in the pre-heated oven for about 25 minutes, until golden brown.

■ Leave the muffins to cool in the pan for 5 minutes, then dust them with a little confectioners' sugar to serve.

Mixed red berry jam

Serves 4

10 oz (300 g) mixed berries (e.g. strawberries, raspberries, blackberries, jostaberries, bilberries, currants)
1 vanilla bean
1 cup (250 ml) cherry or currant juice
5 tbsp (70 g) sugar
grated zest of 1 unwaxed lemon
1 cinnamon stick
1–1½ tbsp cornstarch

■ Pick through the berries, rinse them carefully, and pat dry. Cut any larger berries into halves or quarters, depending on their size.

■ Slit the vanilla bean in half lengthwise and scrape out the seeds. Bring the juice to a boil with the sugar, vanilla bean, and vanilla seeds. Add the lemon zest and cinnamon stick, and simmer over low heat for about 5 minutes.

■ Mix the cornstarch with a little water until it has dissolved. Pour into the boiling juice, stirring vigorously, and leave to simmer for about 3 minutes until the mixture thickens.

■ Remove from the heat, add the fruit, and fold in. Chill for several hours or overnight.

■ Remove the cinnamon stick and vanilla bean before serving. Goes well with whipped cream or vanilla ice cream.

Raspberries—a heavenly snack

The raspberry (*Rubus idaeus*) has had an impressive career to date: from the unassuming, sour little fruit that spent its time at the edge of the forest to a widely appreciated, aromatic berry that no garden should be without. Early observers noted that hinds were especially fond of the berry, so it was long known as the hindberry. Its aroma is reflected in its other common names in other languages, such as "honey berry," while "heaven berry" indicates the heights to which it can grow—in sunny spots on the edge of wooded areas, the bushes can grow to nearly 7 feet (2 m) in height.

As far back as the Stone Age, people were flocking to gather these red berries of the forest as the corn was ripening in late summer. Later adopted by the Romans, raspberries were grown in monastery gardens from the Middle Ages, and soon became an established feature in the community gardens of the 19th century. In addition to the red varieties, yellow raspberries have been available since 1588, these being seen as a particular delicacy at the time. Even black and violet raspberries have been on the market since 1832.

Similarly to wild strawberries, wild forest raspberries, widespread throughout Europe, have an especially intense aroma and contain valuable ingredients. Harvesting the wild fruit can be arduous, however. The thorny, head-high canes form a thicket and often flourish in steep, inaccessible places.

Harvested when fully ripe, raspberries are among the most delicious of berry delights. Optimum ripeness can be recognized by the berries coming away from the flower receptacle, the inner cone, very easily. These pressure-sensitive and perishable berries are delivered as quickly as possible by air to the sheiks of Arabia. For us, they are somewhat closer to hand, fortunately. Here in Europe the berries find their way directly to the kitchen, where they are made into all kinds of culinary delights: jams, jellies, compote, fruit sauces, syrups, vinegar, juice, wine, milkshakes, and all manner of desserts such as soufflés, cold puddings, ice cream, and sorbet. They also largely retain their aroma and intense red color after processing. Most of these delicious berries are enjoyed just as they are, though, directly from hand to mouth, with yogurt, cream, or ice cream, as cake toppings, or—highly recommended—in salads.

I personally enjoy raspberries the most in alcohol. A raspberry liqueur is for me the most noble means of preserving the delicate aroma: pure and ice cold, poured over a serving of vanilla ice cream or to flavor a fruit sorbet, this spirit is a wonderful summer delight. The berries contain, inter alia, sugar and fruit acids, Vitamin C, potassium, iron, magnesium, mucins, anthocyanin, and other flavonoids. In folk medicine, raspberry juice is used against fever and to strengthen the heart.

Young raspberry leaves gathered in the spring are also edible. Dried, they make an aromatic infusion, similar in taste to black tea. Raspberry tea is used as a treatment for diarrhea, and also for gargling in the case of mouth or throat infections. It has diverse benefits during pregnancy and childbirth.

Raspberries are still conspicuous by their absence in my garden, so I will be preparing a bed for them soon. To this end, I will first loosen the earth and add generous amounts of humus. Working in some bark mulch will make the pH value slightly acidic, which helps to prevent root decay in the event of waterlogging as well as the stress-related, highly infectious cane blight. A mulch layer around the plant prevents it drying out when the fruit are ripening and encourages regrowth. As in their natural environment, the location should not be in full sun, otherwise the leaves and fruit can suffer sunburn. The older canes are best cut off close to the ground directly after the harvest. A stabilizing frame ensures that the canes do not get too tangled and that the berries can ripen properly. Most widespread are the conventional summer raspberries and some repeat bearing varieties. As I still want to be able to harvest raspberries later in the year, I will be choosing the more robust autumn raspberry varieties. Their later ripening makes them less susceptible to cane blight. An organic, slow-working berry fertilizer allows the plants to grow evenly and prevents parts of the plant from becoming too soft.

Raspberry bushes develop a great many runners and regenerate themselves in this way. Nevertheless, I would not plant any shoots given to me in my garden. Unfortunately, with neighborly donations over the garden fence, your raspberries can quickly end up with the dreaded cane blight or even stunted growth. I therefore recommend planting healthy, virus-free bushes from a good fruit nursery.

Flavorsome cooking with raspberries

Mojito with raspberries and mint

Makes 1 drink

¼ unwaxed lime
6 raspberries
1 tsp raw sugar
6 mint leaves
2 ½ tbsp (40 ml) white rum
1 tbsp raspberry syrup
3 tbsp crushed ice
scant ½ cup (100 ml) mineral water

■ Slice the lime into thin wedges. Carefully rinse the raspberries and place in a lowball glass together with the sugar and mint leaves. Crush slightly with a pestle.

■ Add the rum and the raspberry syrup and then the ice. Top up with mineral water and serve.

Swiss roll with raspberries

Makes 1 Swiss roll

For the sponge cake
5 eggs
9 tbsp (125 g) sugar
pinch of salt
2/3 cup (100 g) all-purpose flour
1/3 cup (50 g) cornstarch
confectioners' sugar, for dusting

For the filling
4 gelatin leaves
generous 1 lb (500 g) raspberries
7 oz (200 g) mascarpone
7 oz (200 g) quark
1/3 cup (75 g) sugar
2 tbsp lemon juice
2½ tbsp (40 ml) raspberry liqueur
gen. ¾ cup (200 ml) whipping cream
confectioners' sugar

■ Pre-heat the oven to 390 °F (200 °C). For the sponge cake: separate the eggs. Place the egg yolks in a mixing bowl, add about two-thirds of the sugar, and beat with a handheld whisk until light and creamy.

■ Beat the eggs whites with the salt until stiff, slowly adding the rest of the sugar while beating. Fold the beaten egg whites into the egg yolk mixture.

■ Sieve the flour and the cornstarch over the egg mixture, and carefully fold in using a dough scraper. Spread the mixture over a baking sheet lined with wax paper and bake in the center of the pre-heated oven for 10–12 minutes.

■ Sprinkle a large, clean dish towel with sugar and carefully tip the hot sponge cake on to it. Brush the wax paper with a little cold water and remove from the base of the cake. Roll up the sponge cake immediately with the help of the dish towel, and leave to cool.

■ For the filling: soak the gelatin in cold water. Pick through the raspberries, wash them, and pat dry.

■ Mix the mascarpone with the quark, sugar, and lemon juice until smooth. Squeeze out the gelatin, mix with the raspberry liqueur, and heat gently in a saucepan until the gelatin has dissolved.

■ Fold 3 tablespoons of the mascarpone cream into the raspberries then gently stir in the rest. Beat the whipping cream until stiff and fold into the raspberry mixture, finally adding the warm liqueur.

■ Carefully unroll the cooled Swiss roll and spread with the raspberry cream filling. Use the dish towel to roll it up again, cover, and chill for at least 3 hours. Dust with confectioners' sugar and slice.

Variations

This Swiss roll can be made using the same quantity of strawberries or other fresh fruit in season.
Instead of quark and mascarpone, you could use 2 cups (500 g) of whipping cream, or 1½ cups (400 g) of yogurt plus ¾ cup (200 g) of whipping cream—for the lower calorie version.

Chocolate cupcakes with raspberries

MAKES 12 CUPCAKES

For the batter
2½ cups (200 g) raspberries
3½ oz (100 g) dark couverture chocolate
7 tbsp (100 g) butter
2/3 cup (150 ml) milk
1 egg
6 tbsp (100 g) sour cream
7 tbsp (100 g) sugar
1¾ cups (250 g) all-purpose flour
2 tsp cornstarch
1 tbsp baking powder
2 tbsp cocoa powder
pinch of salt

For the glaze
11 oz (300 g) milk couverture chocolate
10 tbsp (150 ml) whipping cream
2½ tbsp (40 g) butter
1 tbsp raspberry syrup

■ Pre-heat the oven to 350 °F (180 °C) (fan). Carefully rinse the raspberries and drain well.

■ Roughly chop the dark couverture chocolate. Place the butter in a saucepan, add the chocolate, and melt over low heat. Remove the saucepan from the heat.

■ Beat together the milk, egg, sour cream, and sugar using a handheld whisk.

■ Combine the flour, cornstarch, baking powder, cocoa, and salt in a bowl and mix together. Add the milk mixture and the chocolate-butter mixture, and stir until well mixed. Fold in the strawberries.

■ Line a muffin pan with paper cases and fill with the cake batter. Bake in the center of the oven for about 30 minutes.

■ Remove the cupcakes from the pan and leave to cool. Then remove the paper cases.

■ To make the glaze: finely chop the milk couverture chocolate and place in a bowl over hot water. Whip the cream, add to the couverture in the bowl, and stir gently until the chocolate has dissolved.

■ Add the butter in flakes and beat in. Stir in the syrup and leave to cool.

■ Stir through thoroughly again and place in a pastry bag with a smooth nozzle. Pipe the cream over the cupcakes and serve.

Japanese wineberry—tasty, yet unknown

Astoundingly, these eye-catching, shiny, red berries had already found their way into European and North American gardens as early as the 19th century. Yet the Japanese wineberry (*Rubus phoenicolasius*) is practically unknown to many. The tart berries are healthy and taste fantastic, the yield is mostly generous, the plants are winter hardy—and they clearly deserve more attention!

At a first glance over the backyard fence, you might think they are blackberry plants. But on closer inspection, the fruit looks more like a raspberry. Japanese wineberries are related to both, though there are significant differences. Their imposing height (up to 10 feet / 3 m) gives them a dominant appearance. The red canes have red thorns. Their greenish-yellow leaves have silvery-white hairs on the underside and straddle the overhanging canes. The exotic-looking flower buds are a sight for any garden. The best is still to come, however—the plump, shiny, orange-red berries, which look like they have been varnished. They are about the size of a raspberry and ripen between mid-July and early September. When fully ripe, the juicy berries come away from the inner cone easily. With a fine, fruity, pleasantly sweet-and-sour flavor, they are reminiscent of kiwis or grapes. A single bush produces up to 9 pounds (4 kg) of fruit in one season.

The berries are delicious fresh and in the kitchen can be used just like raspberries. There is no limit to the creative uses to which they can be put. Wineberries are ideal for use in juice, liqueurs, jams, jellies, and cakes. They are also delicious candied or in a chocolate fondue. They serve as an eye-catching and flavorsome garnish for ice cream, cupcakes, sweet cream desserts, fruit salads, or cheese platters. The fresh, red juice is reminiscent of wine. Wineberries contain vitamins A, B, and C as well as natural pectin and fruit acids. Soft setting jams can be improved through the use of wineberries. These low-calorie berries are also full of minerals such as iron, potassium, calcium, magnesium, and phosphorus. Their anthocyanins function as antioxidants and boost the immune system.

In the garden, a humus soil and sunny to semi-shaded location is advantageous. The berries will not be able to ripen if the soil dries out too much in the summer. A thick mulch layer of compost will provide moisture and humus. The plant is self-pollinating: one bush is enough to yield a good crop of berries. The arched, downward-hanging canes are best trained along a trellis or frame. This helps the fruit to ripen and makes harvesting much easier. Lily of the Valley is ideal for planting beneath wineberries, as it is said to increase their fertility.

Wineberries are frost hardy and need no winter protection. Overhanging canes quickly put out roots into the soil and this, together with the seeds, can mean that your wineberry plants soon make their way into a neighbor's yard. You do need to keep them in check, therefore. Wineberry canes only bear blossoms and fruit in the second year. You therefore need to cut out the harvested canes to ensure regrowth and remove any long, unwieldy ones as required. Infestation with pests such as maggots is rare because the berries are enclosed in sticky sepals until ripe, which affords them natural protection.

May berries—a fine spring treat

The only type of fragrant honeysuckle in central Europe whose fruit can be eaten is the blue-berried honeysuckle (*Lonicera caerulea*). The berries appear particularly early in the year; they are tasty, and healthy. In some regions, however, there are specimens of this now endangered variety that can cause nausea when consumed. It was for this reason that a search was undertaken and—following the fall of the Iron Curtain—May berries (*Lonicera kamtschatica*), a closely related, edible variety, were discovered in Eastern Siberia. The blue fruit is also called the honey berry and in Germany is ripe for picking in May. Each berry measures ½–1 inch (1–2 cm) in length, with an attractive, cylindrical shape. They are a delight to eat freshly picked: May berries are my particular recommendation for innovative cooks and gardeners.

In terms of taste, the modern varieties can be compared to bilberries. The more sunlight the berries enjoy while ripening, the sweeter they will be. In the kitchen, they can be used just like bilberries—the recipes are interchangeable. Should there be enough berries left over after enjoying them fresh, they are certainly worth a try as juice, jam, purée, or compote. The pectin they contain ensures a good set. They also contain abundant Vitamin C, B vitamins, and a high concentration of phytochemicals.

The fragrant, yellow-white blossoms appear in March and ripen into berries within a few weeks. A sunny location is vital, if they are to become sweet. As the earliest of the berries, the May berry is a low-maintenance alternative to bilberries (*Vaccinium corymbosum* or *Vaccinium myrtillus*). They need an acidic soil (pH value: 4–5), which in some locations means preparing a special soil mix

containing peat. The May berry is otherwise very undemanding and can also cope with an average, not too chalky soil. They do need sufficient water, however, so that the leaves do not turn yellow in the summer and thereby compromise next year's harvest. You could place coconut fibers in the planting hole as an added extra for better moisture retention. Organic berry fertilizers and a covering of bark mulch ensure even growth and the appropriate pH value. Coming from Eastern Siberia, May berries are entirely winter hardy and are spared most plant diseases.

With their modest demands and compact size (up to 5 feet/150 cm) May berries are the ideal candidates for balcony gardeners. Well-structured organic potting soil and coconut fibers in the container will produce outstanding growth. In the garden, planting wild strawberries or cranberries beneath the May berries will complete the picture of perfection—these plants have the same soil, care, and location requirements.

Flavorsome cooking with May berries

May berry cookies

Makes 20–24 cookies

10 tbsp (140 g) butter
scant 1 cup (200 g) raw sugar
7 tbsp (50 g) confectioners' sugar
2 eggs
1 vanilla bean
2½ cups (350 g) all-purpose flour
1 tsp baking powder
pinch of salt
5½ oz (150 g) May berries

■ Pre-heat the oven to 350 °F (180 °C). Beat the butter with the raw sugar and confectioners' sugar until creamy, and then fold in the eggs.

■ Slit the vanilla bean open lengthwise, and scrape out the seeds. Combine the flour with the baking powder, salt, and vanilla seeds, and then stir into the egg and butter mixture.

■ Pick through the May berries, wash them, pat dry, and fold into the mixture. Use a tablespoon to place small amounts of the cookie dough on a baking sheet lined with wax paper, leaving some space between each one.

■ Bake in the pre-heated oven for 10–12 minutes. Leave to cool on a wire rack.

Sweet May berry rolls

MAKES 1 LOAF PAN (10 × 10 IN / 25 × 25 CM)

For the dough
3½ cups (500 g) all-purpose flour
pinch of salt
1 cube or sachet active dry yeast
7 tbsp (100 ml) lukewarm water
3½ tbsp (50 g) sugar
1 egg
7 tbsp (100 ml) lukewarm milk
⅓ cup (75 g) butter
flour, for rolling

For the filling
7 oz (200 g) May berries
3 tbsp (45 g) butter
⅓ cup (75 g) brown sugar
⅔ cup (75 g) chopped hazelnuts

For the glaze
3½ tbsp (50 g) butter
⅓ cup (75 g) brown sugar
3½ tbsp (50 ml) milk

■ For the bread dough: place the flour and salt in a bowl and make a hollow in the center. Crumble the yeast into the hollow and then add the lukewarm water, followed by the sugar, and stir.

■ Add the egg, the lukewarm milk, and the butter and then work into an elastic dough, either by hand or using the dough hook of a food processor. Cover the dough and leave to rise in a warm place for about 1 hour, until it has doubled in volume.

■ Knead the dough again, cover, and leave to rise again for a further 30 minutes.

■ Knead again on a floured surface and then roll out into a rectangle (about 16 x 10 inches / 40 x 25 cm).

■ Wash the May berries and pat them dry. Melt the butter and brush over the surface of the dough. Sprinkle with the brown sugar, May berries, and hazelnuts.

■ Roll up the dough from the long edge. Then use a sharp knife to cut the roll into 9 slices about 1½ inches (4 cm) in thickness.

■ Grease a square baking pan and place the rolls in it, leaving some space between each one. Cover and leave to rise again for about 45 minutes.

■ Pre-heat the oven to 350 °F (180 °C) (fan). Melt the butter with the sugar and milk until the sugar has dissolved. Use to brush the rolls and bake in the pre-heated oven for about 30 minutes until golden brown. Cover with aluminum foil, should they become too dark during baking.

■ Remove from the oven and leave to cool on a wire rack. Serve either warm or cold.

Ornamental splendor—the chokeberry

The chokeberry (genus *Aronia*) is the discovery of recent years for me. This plant has great gardening appeal because, as the so-called "four seasons" bush, it always has something to look at. In the spring, its juicy, round, green leaves burst out of red-brown buds. In the early summer, the plant is crowned with clusters of white blossoms. The subsequent abundance of wonderful, black-violet berries then has you looking forward to the fall harvest. To conclude the gardening year, the leaves then turn a bright red-orange, reminiscent of an Indian summer in the plant's North American home. Like the apple, the chokeberry belongs to the rose family (Rosaceae).

The small, apple-shaped berries have an impressive array of extraordinary health benefits. There is little or no documented evidence of their effectiveness, but the chokeberry has long been used in Russian folk medicine. Becoming increasingly significant as a medicinal plant elsewhere in the world, it is said to help with high blood pressure, circulatory disease, and arteriosclerosis. The fruit is full of healthy ingredients that boost the body's defenses and assist in warding off disease. The chokeberry is in a league of its own among other garden berries, with its unique composition of active agents. Worthy of mention are the phytochemicals, particularly the deep violet coloring agents (anthocyanin). These are present in abundance in the chokeberry, while especially effective are the flavonoids that react in human metabolism

with damaging free radicals, binding chemically with them and thus having a cell-protecting, antioxidant effect. This benefits the body in times of stress and illness in particular. Also, the amount of health-boosting polyphenols is some five times higher than in other garden berries. Among the many vital nutrients contained in this fruit are the vitamins C, E, and K, beta-carotene, folic acid, iron, and potassium.

The deep violet berry juice is the easiest way to access all of the healthy active agents. You can make this "health elixir" yourself using the classic process of gentle juice extraction. The refreshingly bitter aroma and appealing deep violet color of these berries can be used to enhance all kinds of culinary delights, and also as a colorant for sauces, jams, and sweet, milk-based dishes, among others. Try them out the next time you are making strawberry jam! Chokeberries are also highly recommended as a valuable dried fruit, well suited to baking. You can also preserve the fruit in alcohol and the berries are even suitable for candying.

In the garden, the undemanding chokeberry is ideally suited to the "lazy bed" method of gardening. The soil should be low in lime and not too alkaline (pH value: 6–6.5). Organic berry fertilizer and watering at the height of summer improve the development of the fruit. The chokeberry prefers sunny spots, but can also cope with a semi-shaded location—but the berries do develop significantly more sugar in a sunny location.

When the Russian grower Mitschurin took on this berry for fruit cultivation in around 1910, it was the plant's frost hardiness that appealed to him above all (to –9.4 °F / –23 °C). This makes it suitable for growing in less climatically ideal locations. The most widespread is the *Aronia melanocarpa*, the black chokeberry. Its cultivars differ from the wild variety largely in the size of the berries. The 1970s saw the start of chokeberry cultivation in the German Democratic Republic and in the German state of Saxony there are still around 99 acres / 40 hectares under cultivation.

The red chokeberry (*Aronia arbutifolia*) has become an established feature in my garden. Its red berries do not match the flavor of the black variety, but its high yield provides a sound basis for all manner of wild berry dishes. I am particularly fond of the chokeberry as a shrub, and have planted it along my terrace so that I can pick the berries at my leisure and enjoy the changing color display with the seasons.

Flavorsome cooking with chokeberries

Chokeberry milkshake

Serves 4

7 oz (200 g) chokeberries
5 oz (150 g) bilberries or blueberries
dash of lime juice
1²/₃ cups (400 ml) chilled coconut milk
1 cup (250 ml) chilled milk
3–4 tbsp honey, to taste

■ Pick any leaves or stalks from the chokeberries and bilberries, wash the fruit under running water, and drain well.

■ Combine with the lime juice, coconut milk, milk, and honey in a blender and purée until creamy. Pour into glasses and serve.

Variations

The milkshake also tastes good when made with chokeberries alone.

The coconut milk can be replaced by the corresponding amount of full cream milk. If you want an especially creamy milkshake, combine 2 cups (500 ml) of full cream milk with 10 tablespoons (150 g) of yogurt.

Honey or sugar can be reduced, or left out entirely in favor of a ripe banana.

If calories are not an issue, you can crown your milkshake with a scoop of vanilla ice cream.

Chokeberry juice

■ Wash and drain the chokeberries. Place them in a saucepan with the lemon juice and sugar, and heat. Add the water and simmer over gentle heat for about 20–25 minutes.

■ Pass the liquid through a strainer or muslin cloth, squeezing the berries well.

■ Re-heat the juice and boil for a further 15 minutes.

■ Pour into clean, sterilized bottles, seal well, and leave to cool.

MAKES 2 × 8-OZ (250-ML) BOTTLES

2¼ lb (1 kg) chokeberries
juice of 1 unwaxed lemon
scant 2 cups (400 g) sugar
2½ cups (600 ml) water

Flowering quince—a treat in every sense

The abundant blossoms of the flowering quince stand out in the hedgerow from about April—in orange, white, pink, or coral red. The yellow fruit then appear between the thorny branches in October. That is when the question generally arises—what is that fruit, which feels so crisp and smells so wonderful?

The flowering quince is a delight, not just for the eye, but also for the nose and palate. The fruit is full of essential oils, even being used in perfume manufacture. It quickly scents its surrounding area, from cupboard to entire room. In the kitchen, it is used in just the same way as true quince (*Cydonia oblonga*). The flowering quince (*Chaenomeles* sp.) is also known as the Japanese quince, or wild quince. I personally prefer flowering quince to true quince, because the harvest from its bushes is much greater than that from an often expansive quince tree.

The decorative flowering quince originated in Asia, where it was widespread in China, Japan, and Korea in particular, being planted in the gardens of nobility as a status symbol. "Plant hunters" brought it to Europe as a novelty toward the end of the 19th century and it was subsequently found as an ornamental shrub in many gardens. Its career as a wild fruit only took off later. It was the well-known variety "Cido," its fruit containing more Vitamin C than lemons, which was introduced to the market as the "Northern Lemon" by a resourceful gardener in the 1970s.

The fruit are initially very hard and pressure-sensitive, and come away from the branch with some resistance. But if you wait until the first frost, the cell structure softens and the fruit becomes easier to work with. You peel the flowering quince, and remove the core as with an apple. The fruit is inedible raw and it is only once processed that it develops its fantastic aroma. Typical recipes include jams, pastes, compote, purée, juice, syrup, wine, or liqueur. A delicious, low-calorie, and very healthy alternative to jelly babies is quince leather, for which the fruit purée is dried in the oven as confectionery (see Sea buckthorn leather, p. 125). Also relevant for the regionally-minded gardener-gourmet is the use of sour quince juice as an alternative or supplement for lemon juice. Healthy, refreshing "lemonade" can also be made from flowering quince, entirely without the use of the citrus fruit. The natural sugar content of the former is some 50 percent less

than that of other fruit. Flowering quince contains about three times as much pectin as apples and is therefore ideal as a setting agent. It is also used as a traditional medicine for diarrhea.

Growing up to 5 feet (1.5 m) in height, these undemanding shrubs flourish best in a deep soil that is not too chalky. Flowering quince is often planted in wild fruit hedges, where its thorns keep unwanted guests at bay. The plant is highly amenable to pruning and so can also be trained as a border or shaped hedge, thus giving its abundant blossoms even greater prominence.

Flowering quince is so winter-hardy it can be grown in Latvia and even further east. Its early blossoming in the spring means that planting in early fall is important. This enables the plant to take root properly in its new location. Should blossoms be lost due to late frosts, the flowering quince will simply blossom again. Two different varieties of this decorative shrub are ideal for efficient pollination and a good harvest.

Flavorsome cooking with flowering quince

Lentils with vegetables and quince

Serves 4

7 oz (200 g) dried lentils
9 flowering quinces
2 tbsp lemon juice
3 carrots
1 parsley root
11 oz (300 g) celeriac
1 tbsp rapeseed oil
4 cups (1 liter) vegetable stock
10 tbsp (150 ml) cream
3 tbsp clarified butter
2–3 tbsp balsamic vinegar
salt and freshly ground black pepper

■ Wash the lentils thoroughly under running water. Heat sufficient water in a saucepan and boil the lentils, without salt, for 20–25 minutes. Then drain them and rinse under cold water.

■ Rub the quinces with a cloth, wash and peel them, remove the cores, and cut into thin wedges. Drizzle with lemon juice.

■ Peel the carrots, parsley root, and celeriac and cut into pieces about ½ inch (1 cm) in size. Sauté the vegetables in hot oil for 2–3 minutes, stirring all the time.

■ Pour in the vegetable stock and cream, and simmer for about 10 minutes over medium heat until the vegetables are cooked. In the meantime, heat the clarified butter and cook the quinces, covered, over low heat for about 15 minutes.

■ Add the lentils to the vegetable mixture and simmer for a further 5 minutes. Season with balsamic vinegar, salt, and pepper, and spoon the portions onto plates. Garnish with quince wedges before serving immediately.

Venison meatloaf with quince

Serves 6

2 stale bread rolls
2 onions
1½ tbsp (20 g) butter
4 cups (500 g) pre-cooked, peeled chestnuts
1¼ lb (600 g) leg of venison, boned
2 eggs
7 oz (200 g) minced veal
2 tbsp quince jelly
2 tbsp fresh parsley
salt and freshly ground black pepper
1–2 tsp venison spice mix
2½ tbsp (40 g) clarified butter

■ Pre-heat the oven to 390 °F (200 °C). Soften the bread rolls in lukewarm water.

■ Peel the onions, dice them finely, and sauté in the butter. Then leave to cool slightly. Roughly dice the chestnuts.

■ Wash the venison, pat dry, dice roughly, and place in a large bowl. Add the eggs, the sautéed onions, minced veal, the well squeezed out bread rolls, quince jelly, and the chopped parsley.

■ Mix all the ingedients together well and then fold in the chopped chestnuts. Season with salt, pepper, and venison spices.

■ Shape the mixture into a loaf (about 6 x 8 inches / 15 x 20 cm) and place in a roasting pan, together with the clarified butter. Bake in the oven for 50–60 minutes until crispy.

■ Remove from the oven and slice. Serve with boiled potatoes, cranberries, and a cream sauce if preferred.

Figs—the divine fruit

If Eve had only known what snacking on the forbidden fruit would lead to… To my mind, however, the wonderful fig tree definitely belongs in fruit paradise.

The common fig (*Ficus carica*) is one of mankind's oldest cultivated plants, being grown in the Mediterranean region since Antiquity. At that time, figs were a staple foodstuff and, when dried, were an important provision for travelers. The Romans, too, brought the sweet fruit back home from their conquests abroad.

Figs are still cultivated as trees in Mediterranean countries and are often to be found growing wild. There are now also varieties that ripen well in central Europe and which have a certain degree of frost hardiness. Cultivation is especially successful in wine-growing areas. In one part of Germany, there is even a "fig exchange" where fig fans are able to swap information and, more especially, figs themselves.

Pure flavor—sliced, fresh figs with a scoop of vanilla ice cream and a little light cream is a simple, but wonderful, treat. In the summer, fresh, grilled figs are delicious with goat's cheese or grilled steak from the barbeque. The combination of savory and sweet is the taste secret here. Other flavorsome combinations include lavender, Gorgonzola, or honey. Unfortunately, the fresh fruit keep for only a short while in the refrigerator. Preserving them whole in alcohol or making them into fig mustard, a purée, or fig jelly, though, means that you can enjoy this summer fruit through fall to the winter. The blue-violet examples are said to be tastier than their green sisters, probably because their color means they absorb the sun's warmth more quickly and so form more sugar.

Figs comprise 80 percent water and have a great deal of fructose; they are sugar sweet, and yet low in calories. Their natural sweetness makes fresh figs in particular a healthy alternative to chocolate (about 30 kcal per fig). Minerals such as iron, potassium, magnesium, and phosphorus supplement the high dietary fiber content, which gently stimulates the digestive system. The magnesium also has a stress-reducing effect. Figs are a healthy "brain food" and are also good for people who eat a lot of meat, sausages, and sugar—they have the highest alkaline values of all foodstuffs and neutralize acid-forming foods. Figs available commercially have often been subjected to chemical treatment and so it is best to buy the sulfur-free and/or untreated dried fruit from organic suppliers.

In the garden, fig trees need an especially warm, wind-protected spot; mine are located directly in front of a screen wall—even here, in Central Germany, I am able to enjoy a fig or two. Just as important for a sure harvest is the variety of plant. I have chosen a Swiss fig variety, while there are other suitable varieties from France and the Mediterranean countries (see Recommended varieties, p. 180). Particularly in the early years after planting, the trees should be wrapped in white garden fleece in the winter. The prospect of sweet treats in the summer makes it worth the effort.

Flavorsome cooking with figs

Baked figs with Gorgonzola and honey sauce

Serves 4

olive oil, for greasing
4 fresh figs
3½ oz (100 g) Gorgonzola
2 tbsp honey
2 tbsp balsamic vinegar
salt and freshly ground black pepper

■ Pre-heat the oven to 430 °F (220 °C).

■ Brush four small, ovenproof dishes with olive oil.

■ Cut the figs into three and place in the dishes. Crumble the Gorgonzola over them.

■ Bake in the top part of the oven for about 5 minutes.

■ In the meantime, combine the honey and the vinegar. Season with salt and pepper. Drizzle over the figs and serve immediately.

Chicory salad with pancetta and grilled figs

SERVES 4

3 heads of chicory
1 small head of radicchio
8 ripe figs
7 oz (200 g) pancetta
a little fat, for greasing
4 tbsp olive oil
salt and freshly ground black pepper
2 tbsp acacia honey
1 tbsp lemon juice

■ Remove the outer leaves from the chicory, wash, and pat dry. Then cut the chicory heads in half, remove the hard stalk end, and cut into strips. Separate the radicchio leaves, wash, dry in a salad spinner, and cut into strips.

■ Wash the figs and cut them in half. Slice the pancetta into narrow strips and dry-fry in a hot skillet until crispy. Remove and drain on paper towels. Heat the oven broiler.

■ Place the figs, sliced side up, on a greased baking sheet and drizzle with 1–2 tablespoons of olive oil. Season with salt and pepper, and grill for about 5 minutes.

■ During the cooking time, drizzle the figs with 1 tablespoon of honey and turn them once. Combine the honey with the lemon juice and the rest of the olive oil.

■ Combine the chicory and the radicchio and place on individual plates. Arrange the figs on top and sprinkle with pancetta strips. Drizzle with honey and season to taste with salt and pepper.

Bilberries—from the garden and the woods

Wild bilberries, or blueberries: they remind me of the carefree childhood days when I used to visit my grandparents in Germany's Eifel region. Together with my cousins, we children would comb the woods on hot summer days, and everywhere we would find the bright blue berries—a little paradise. Our fingers and tongues were soon colored blue. Back at my grandparents' house came our next serving, crowned with fresh cream.

In European regions, the bilberry (*Vaccinium myrtillus*) often grows at the edges of woods, in acidic soils. The evergreen bushes have no thorns, which makes picking easier. The plants reach no more than about 16 inches (40 cm) in height, so they are ideal for little hands. Or even for very small hands—it is said that forest gnomes and their friends frequent bilberry bushes…

Bilberries are among the most popular of wild fruit. As with mushroom gathering, there are still families today who head off into the woods with a basket in hand. In the 1980s, efforts began to cultivate regional wild plants in order to fulfill the large demand for these little berries. Particularly successful was the cultivation of the blueberry (*Vaccinium corymbosum*), which, like huckleberry, is also native to North America. This yields 9–18 pounds (4–8 kg) of the sought-after fruit per bush. They leave no evidence of snack attacks, though, because the blue color is restricted to their skin. The fruit flesh is a pale gray-blue and does

not transmit color during processing. Harvesting this cultivated variety of blueberry, known as high-bush or swamp blueberry, also turned out to be easier on people's backs than the wild, low-bush variety, as it grows up to 40 inches (1 m) tall. This blueberry as cultivated in Europe is a national fruit in the USA where it has been grown on large plantations since 1900.

Bilberries are very healthy and have healing effects. Fresh, they stimulate the digestion. Dried bilberries, however, are a gentle home remedy for diarrhea, especially popular with children. Gargling with warm bilberry juice is recommended for mouth and throat infections. Eating a handful of the fruit daily provides measurable support for memory and learning capabilities well into old age, according to the results of a study by the University of Boston. The anthocyanins contained in the blue color also absorb the damaging free radicals in our bodies, protect the blood vessels, and help to prevent heart and circulatory disease.

They also contain notable amounts of vitamins A, C, E, and various B vitamins, as well as the minerals calcium, iron, potassium, magnesium, sodium, phosphorus, and zinc. Despite this wealth of ingredients, bilberries are very low in calories (37 kcal per 3½ ounces / 100 g). They have a moisture component upward of 80 percent—a wonderful berry for healthy, slimming, berry delights! The fruit does not keep for long, and so the berries are used for processing in numerous ways. They freeze very well when freshly picked. They retain their shape in cakes and gâteaux, making them visually appealing. Pancakes with blueberries and maple syrup are just one delicious way of serving the fruit.

The tapeworm risk when gathering edible wild fruit close to the ground has been downplayed significantly by experts in recent years, but of course you should always wash wild berries thoroughly before eating them. If you want to be 100 percent certain, however, you should heat the fruit before eating it (see p. 18).

For the gardener, both the bilberry and blueberry varieties with their white or pink blossoms, dark blue fruit, and fiery red leaves in the fall are sought-after bushes. As plants of the woods and moors, however, they require the right home, with a consistently acidic soil. Add a mixture of coniferous wood shavings, bark pieces, oak leaves, and sulfur to the soil in the planting hole (see p. 26). An organic rhododendron fertilizer will easily meet the plant's low nutrient requirements. Bilberries do need adequate water supplies, however, particularly during the summer months. They are self-pollinators, so one plant is sufficient to secure a harvest. When grown in containers on the balcony and the terrace, these berry bushes make a pretty sight almost year round!

Harvest the fruit by picking or, if they are your own bushes, by using a berry comb. However, the latter method pulls off unripe berries as well as a great many leaves, which impairs the plant's growth and therefore next year's harvest. As a result, the use of a berry comb is prohibited for the harvesting of wild bilberries in many areas. I am happy to take the time to pick my bilberries individually.

For Recommended varieties, see p. 179.

Bilberry yogurt ice cream

SERVES 4

11 oz (300 g) bilberries
2 tbsp (30 g) honey
1–2 tbsp lemon juice
1¼ cups (300 g) yogurt
10 tbsp (150 ml) whipping cream

■ Remove any leaves or stalks from the bilberries, wash the fruit, and pat dry. Place in a blender with the honey and lemon juice, and purée. Pass through a sieve afterward, if desired. Then fold in the yogurt.

■ Beat the cream until stiff and fold into the berry mixture.

■ Either freeze in an ice cream maker or place in a shallow metal container and freeze for at least 4 hours, stirring well every 30 minutes.

■ Use an ice cream scoop to serve.

Flavorsome cooking with blueberries / bilberries

Blueberry pancake delight

SERVES 4

7 oz (200 g) bilberries
5 tbsp butter
scant 1 cup (120 g) all-purpose flour
1 cup (250 ml) milk
4 eggs
2 tbsp sugar
pinch of salt
confectioners' sugar, for dusting

■ Pick through the blueberries, removing any leaves and stalks, wash the fruit, and drain in a fine mesh sieve.

■ For the pancake: put 2 tablespoons of butter in a small saucepan and melt over low heat. Place the flour in a large bowl, add the milk, and beat with a handheld whisk until smooth.

■ Separate the eggs. Add the egg yolks, melted butter, 1 tablespoon of sugar, and salt to the flour-and-milk mixture, and stir until smooth.

■ Beat the egg whites with 1 tablespoon of sugar until stiff and then fold into the pancake mixture.

■ Melt 1 tablespoon of butter in a non-stick griddle or skillet. Pour the pancake mixture into the griddle, sprinkle with the berries, and cook over medium heat for about 3–4 minutes so that the underside browns slowly. Then turn over and cook on the other side for 2–3 minutes until the pancake is light golden brown in color.

■ Slip the thick pancake onto a chopping board and pull to pieces with two forks.

■ Wipe the griddle with paper towels and melt another 2 tablespoons of butter. Sprinkle a little confectioners' sugar in the griddle and allow to caramelize slightly.

■ Place the pancake pieces in the griddle, turning them in the caramelized sugar until the pieces are well browned on all sides.

SERVES 4

4 slices of pumpernickel
9 oz (250 g) bilberries
11 oz (300 g) salad greens (e.g. arugula, baby spinach)
1 tbsp lemon juice
2 tbsp white balsamic vinegar
4 tbsp rapeseed oil
pinch of sugar
salt and pepper
7 oz (200 g) Gorgonzola

Bilberry and Gorgonzola salad

■ Break or slice the pumpernickel into bite-size pieces, then brown lightly in a hot skillet. Remove from the skillet, and leave to cool.

■ Pick over the bilberries, wash them, and leave to drain well. Wash the salad greens and dry in a salad spinner.

■ For the dressing, combine the lemon juice with the vinegar and oil. Season with sugar, salt, and pepper.

■ Combine the pumpernickel with the salad greens and bilberries, arrange on plates, and crumble the cheese on top. Drizzle with the dressing and serve.

Lingonberries—sweet or savory

People have been gathering these little, red berries since prehistoric times. The lingonberry (genus *Vaccinium*) is a child of the high heaths and moors and so found its ideal living conditions in the post-Ice Age landscape. At that time the small bushes, about 12 inches (30 cm) in height, covered large areas, visible from afar as a bright red-gold carpet in the fall when entire villages would gather to harvest them. Lingonberries contain relatively little water, so could be dried for consumption through the winter as a valuable source of vitamins. Peat exploitation and the conversion from a natural to a man-made environment has meant that lingonberries are now rare in the wild. They are still there to be gathered in high mountainous areas only, or to be admired in highly protected moorland regions.

A very healthy fruit, lingonberries contain several vitamins (C, beta-carotene/A, B1, B2, and B3) as well as valuable minerals like calcium, magnesium, phosphate, and potassium. Medical studies have found evidence of a preventative impact on urinary complaints such as bladder or kidney infections. The regular consumption of lingonberries prevents the build-up of bacteria in these regions, primarily due to the effects of the anthocyanins in the berries' red coloring agents. The fruit is also said to be very helpful against colds and rheumatic conditions, while those who do not get enough exercise in the winter can boost their intestinal flora with these berries, to help regulate intestinal activity.

The classic use for the lingonberry, which is a member of the cranberry family, is in sauces or compotes to accompany venison dishes. The cooked berries also go very well with broiled or grilled meat and fish, or with a cheeseboard. Their refreshingly tart aroma harmonizes well with sweet ingredients, as in cakes or muffins, where the fruit can simply be stirred in and baked. As a topping, ingredient, or filling they work well with shortcrust pastry, sponge cake mixes, and batter-based dishes. They are delicious in combination with chocolate or apples—I really enjoy them in baked apples. In the summer I enjoy a slightly sweetened lingonberry juice mixed with mineral water or apple juice. The fresh berries can be stored for several months in a cool, dark place, as they have a natural wax coating that protects and preserves them.

Growing these valuable fruit bushes in the garden requires an acidic soil (pH value 4–6). To this end you will need to remove the soil from the planting hole and replace it with a mixture of soil, oak leaves, coniferous wood shavings, and sulfur. A thick layer of coniferous bark mulch then needs to be applied after planting. If you do decide to create a "moor bed" like this, then you can plant the lingonberries together with bilberries and cranberries. Azaleas and rhododendrons will also flourish in these conditions. The winter hardiness of lingonberries means that growing them even at higher altitudes has good prospects of success. You should cover the plants with spruce branches in the winter, however, as late frosts can cause leaf and blossom damage. The small white or pink blossoms appear in May and the berries can be harvested in September; some cultivars even bear fruit into late October.

Flavorsome cooking with lingonberries

Couscous with chicken, mushrooms, and lingonberries

Serves 4

9 oz (250 g) couscous
1¼ cups (300 ml) water
salt
1¾ oz (50 g) fresh lingonberries
9 oz (250 g) wild mushrooms (e.g. chanterelles)
1 ripe mango
2 chicken scallops, approx. 7 oz (200 g) each
freshly ground black pepper
4 tbsp vegetable oil
½ cup (125 ml) dry white wine
10 tbsp (150 ml) cream

■ Place the couscous in a bowl. Bring the water, together with the salt, to a boil and pour over the couscous, just covering it. Leave to soak for about 3 minutes and then loosen with a fork.

■ Rinse the lingonberries and pat dry. Set aside a small handful for the garnish and stir the rest into the couscous.

■ Clean the mushrooms and cut them into small pieces. Peel and pit the mango then cut it into thin strips.

■ Wash the chicken scallops, pat dry, cut into strips, and season with salt and pepper. Brown on all sides in a skillet with 2 tablespoons of oil for a few minutes, then remove. Brown the mushrooms with the remaining oil in the same skillet.

■ Add the white wine and cream, stirring occasionally. Add the chicken and warm through, seasoning again with salt and pepper.

■ Arrange the couscous on the plates and the chicken and mushrooms on top. Serve garnished with the mango slices and lingonberries.

Baked apples with lingonberries

SERVES 4

butter for greasing
4 tart apples
3½ oz (100 g) lingonberry jam
4–5 tbsp ground hazelnuts
1–2 tbsp cream
1 tbsp honey
pinch of cinnamon
butter for topping

■ Pre-heat the oven to 350 °F (180 °C). Grease a baking dish or casserole with butter.

■ Wash the apples and remove the cores. Mix together the jam, hazelnuts, cream, honey, and cinnamon.

■ Using a teaspoon, fill the apples with the jam mixture. Dot with butter and bake in the oven for about 30 minutes.

Cranberries—American power berries!

The cranberry (genus *Virburnum*) originated in North America and rose to fame through Thanksgiving. This annual festival, celebrated with a traditional dinner of turkey, pumpkin pie, and cranberry sauce, goes back to the time of the Pilgrim Fathers. It was only with the help of the local Native American people that these early settlers were able to survive. The former were generous with donations of foodstuffs and showed the pilgrims how to prepare the local produce. Out of gratitude, the Pilgrim Fathers held a three day harvest festival for the local people in 1621 and cranberries remain an essential feature of Thanksgiving to this day.

The cherry-size red berries are among the most popular fruit in the USA and are becoming increasingly available commercially in Europe, where the cranberry is already an ingredient in a surprising range of snacks, mueslis, baking mixes, juices, and cocktails. The raw fruit has a tart, acidic flavor, but when cooked develops a milder taste reminiscent of lingonberries. Cranberries are especially good in combination with sweet fruit, or as a sweet-and-sour sauce to accompany venison and poultry. The dried fruit can be used like raisins and works well in all kinds of baked goods.

The vitamin-rich cranberry is considered a secret weapon against all kinds of infections, especially of the urinary tract. There are no scientific studies to date clearly proving the medical effectiveness of cranberries, but the preventative or curative use of cranberry

juice for bladder infections is widespread today among women in particular. The fruit stimulates a flushing out of the bladder and urinary tract, which is said to prevent the build-up of harmful bacteria. The same also applies to dental plaque, which can be significantly reduced with cranberries. Here it is the proanthocyanins, from which the fruit derives its red color, that are especially effective. Cranberries are also very low in sugar and so are ideal for a healthy diet. The plant grows to about 12 inches (30 cm) in height and needs a sunny or semi-shaded location for the fruit to be able to ripen. As with bilberries, the soil needs to be acidic (pH value 4.5–5.5). Mix the soil from the planting hole with coniferous wood shavings or aged coniferous wood bark, oak leaves, and sand. A sprinkling of sulfur will also ensure the long-term lowering of the pH value. Untreated bark mulch ensures a balanced soil climate, with an acidic, organic rhododendron fertilizer being advisable as well. The substrate is easily controlled in a container, which is why cranberries are ideal for growing on balconies and terraces—and even in hanging baskets.

Flavorsome cooking with cranberries

Cranberry sticks

Makes 25–30 sticks

14 tbsp (200 g) soft butter
7 tbsp (100 g) sugar
2 tbsp vanilla sugar
pinch of grated zest from an unwaxed lemon
1 1/3 cups (200 g) all-purpose flour
½ cup (50 g) ground blanched almonds
2/3 cup (80 g) dried cranberries

■ Pre-heat the oven to 350 °F (180 °C).

■ Beat the butter together with the sugar, vanilla sugar, and the lemon zest until creamy.

■ Fold in the flour, almonds, and cranberries and work into light, crumbly dough.

■ Using your hands, shape small portions of the dough into cylinders about 3 inches (8 cm) long and place them on a baking sheet lined with wax paper, flattening them slightly.

■ Bake in the oven for about 15 minutes until pale gold in color. Carefully remove from the baking sheet and leave to cool.

Flammkuchen (Alsatian pizza) with ewe's milk cheese, smoked turkey, and cranberry jelly

Serves 4

For the cranberry jelly
2¼ lb (1 kg) fresh cranberries
a little water
2¼ cups (500 g) gelling (jam) sugar

For the flammkuchen
⅓ oz (10 g) fresh yeast
7 tbsp (100 ml) lukewarm water
1¾ cups (250 g) all-purpose flour
pinch of salt
1 tbsp sunflower oil
gen. ¾ cup (200 g) sour cream
3½ oz (100 g) smoked turkey breast, sliced
5½ oz (150 g) ewe's milk cheese
pepper, oregano

■ For the jelly: wash the cranberries and place in a saucepan with a little water. Cover and simmer for 15 minutes until the fruit are soft. Pass through a sieve or a food mill, if desired, but the jelly is just as good with the fine strips of skin left in.

■ Add the gelling sugar and boil for about 4 minutes before testing for setting.

■ Pour the jelly into clean, sterilized jars. Seal and turn upside down for a few minutes, then leave them to cool the right way up.

■ For the flammkuchen: dissolve the yeast in lukewarm water.

■ Add the flour, salt, and sunflower oil, and knead into smooth, elastic dough. Add a little flour or water as required. Shape into a ball, cover, and leave to rise in a warm place for about 1 hour.

■ Pre-heat the oven to 430 °F (220 °C).

■ Knead the dough again well and divide into four pieces. Roll these out thinly on a floured surface and then place on a baking sheet or pizza stone lined with parchment paper.

■ Spread with the sour cream, leaving the edges free. Layer with the turkey breast slices. Dice the cheese and spread over the dough pieces and then season to taste with pepper and oregano.

■ Bake in the oven for about 10 minutes, until crispy. Serve garnished with the cranberry jelly.

Goji berries—the anti-aging fruit

Who would have thought it? They have been a feature of gardens and parks for decades—and goji berries (*Lycium barbarum*) also flourish along freeways and highways, being planted to stabilize steep slopes. But hardly anyone here in Europe knew them by their Chinese name "goji" until about five years ago. Here, they are simply called "boxthorn" (*Lycium europaeum*), and favor poor, stony, and dry soils in particular. Growing to about 10 feet (3 m) in height, the shrub's arching, hanging branches provide a display of violet blossoms in the early summer. Their appearance reflects their membership of the nightshade family, which has produced a great many edible and medicinal plants.

Goji berries are bright red, tear-shape, and measure about ½ inch (1 cm) in length. They have an intense, refreshing, and slightly tart flavor. Their aroma can vary between sweet and very bitter, depending on the plant, the location, and the degree of ripeness. They can be eaten fresh, dried, cooked, and baked.

Currently the rage among A-listers as the "anti-aging berry," the goji berry is enjoying a major surge in international distribution and acceptance. The health benefits of the fruit have long been known in traditional Chinese medicine. In China, the berries are put to a wide range of uses—as an important tonic for boosting the immune system and the body, protecting against emaciation, and for enhancing overall fitness. The berries are said to improve liver function, blood count, and skin condition, as well as vision. They are used in the treatment of diabetes, impotence, tinnitus, vertigo, deafness, and infertility. Their effectiveness is disputed, but has been evidenced by numerous international studies. The list of phytochemicals is impressive. Worthy of particular mention is the high proportion of amino acids (up to 5 percent), polysaccharides, and coloring agents. Essential oils and terpenes are also present, with the natural steroids and peptides being of particular interest. These active agents lower blood pressure and have an overall stimulatory and "regenerative" effect. Goji berries have an extraordinarily high antioxidant effect.

There is no difference in the health benefits of the fruit, whether it is eaten fresh from the bush or dried. I like goji berries best in muesli, salads, or mixed jams and chutneys. Their aroma harmonizes especially well with other wild berries. They are also great in cupcakes or muffins, remaining juicy yet firm after baking. Tests on conventionally-farmed goji berries have at times revealed increased levels of pesticides, so it is worth purchasing the berries from controlled organic sources or growing your own shrubs. In China, the young shoots of the *Lycium chinensis* variety are steamed as a vegetable. This would be a new culinary use for cuttings if you need to thin out your goji berries.

In the garden, goji berries prefer poorer soils, while the location should be stony and well-drained. Some wind protection is advantageous, to prevent the arched branches from breaking. I grow mine along the fence, fastened loosely. Too many nutrients encourage mildew, which can be combatted with an organic solution. Some varieties do not blossom immediately, but only on two-year-old branches, so choice of shrub is important. Goji berries are self-pollinating, but you will improve the fruit set and harvest by planting a number of bushes and different varieties.

Flavorsome cooking with goji berries

Goji berry cake

Makes 1 springform pan (9 ½ in / 24 cm diameter)

butter and flour for preparing the pan
4 eggs
¾ cup (175 g) soft butter
generous ¾ cup (175 g) sugar
scant 2 cups (175 g) ground almonds
scant 2 cups (175 g) ground hazelnuts
9 tbsp (70 g) dried goji berries
2 tsp baking powder
1 tsp ground cinnamon
pinch of ground cloves
pinch of salt
scant 1 cup (100 g) confectioners' sugar
1–2 tbsp orange juice

■ Pre-heat the oven to 350 °F (180 °C) (fan). Grease the springform pan and dust with flour.

■ Separate the eggs and beat the egg whites until stiff. Beat the egg yolks together with the butter and sugar.

■ Combine the almonds with the hazelnuts, goji berries, baking powder, cinnamon, cloves, and salt. Fold into the egg yolk mixture, followed by the beaten egg whites.

■ Pour into the prepared pan and smooth the surface. Bake in the pre-heated oven for about 40 minutes. Remove from the oven, leave to stand in the pan for a few minutes, and then transfer to a wire rack to cool.

■ Mix the confectioners' sugar with the orange juice to make a sticky glaze, and brush over the cake.

Biscotti with chocolate and goji berries

Makes about 40 biscotti

3 oz (80 g) dark couverture chocolate
2 tbsp (30 g) butter
1¾ cups (250 g) all-purpose flour
1 tsp baking powder
2 tbsp cocoa powder
2 eggs
2 tbsp milk
generous ¾ cup (175 g) sugar
pinch of salt
1⅓ cups (200 g) blanched almonds
10 tbsp (80 g) goji berries
1 egg white
flour for dusting

■ Roughly chop the couverture chocolate and melt together with the butter in a bowl over hot water, or in a bain-marie.

■ Sieve the flour, baking powder, and cocoa into a mixing bowl and create a well in the center. Place the melted chocolate and butter mixture, eggs, milk, sugar, and salt in the well, and work into a smooth dough.

■ Pre-heat the oven to 300 °F (150 °C) (fan). Knead the dough again on a floured surface, gradually working in the almonds and the goji berries.

■ Shape the dough into two rolls (about 2 inches / 4 cm in diameter) and flatten them slightly. Place on a baking sheet lined with wax paper. Lightly whisk the egg white and use to brush the dough rolls. Bake in the pre-heated oven for about 20 minutes.

■ Remove from the oven, allow to cool slightly, and then cut the rolls diagonally into slices about 1 inch (2 cm) thick. Reduce the oven temperature to 212 °F (100 °C) and place the slices, cut side upward, in the oven again to dry for about 10–15 minutes.

The biscotti keep for several months stored in an air-tight container.

Schisandra—five flavors, one berry

The schisandra berry (*Schisandra chinensis*) is something of a newcomer to Europe. It comes from the Far East, where it grows throughout the region from the jungles to the high mountains. The schisandra berry has been used in traditional Chinese medicine for over 2000 years. The leaves, young shoots, and of course the berries are used for teas, as vegetables, and to enrich all kinds of dishes. Here in Europe, the schisandra berry's strong revitalizing effects mean that its profile is rapidly growing. New products, particularly in the form of food supplements, are making their way onto the market. Furthermore, there is plenty to be discovered on the culinary front as well. The red berries have a characteristic flavor all of their own. In China, the plant is called "Wu Wei Zi," literally the "five flavor berry," due to the fact that the berries combine five flavors in one fruit! The fruit flesh and skin are sweet and sour with a salty nuance, while the seeds taste bitter and spicy. This fascinating berry plant is worth discovering!

The schisandra berry enjoys "miracle plant" status—but why? In China, it is a symbol of longevity and is one of the most important toning (invigorating) medicinal plants with a wide range of applications. It is used for both physical and mental complaints. Like ginseng, this medicinal plant is referred to as an "adaptogen," meaning that it has a positive and balancing effect on various illnesses and medical complaints. Longer term consumption is said to make a palpable difference to the human body's ability to adapt to stress and different environmental conditions. In Asia, the red berries are used to treat depression, insomnia, nightmares, mood swings, concentration problems, irritability, and memory problems. On a physical level, its areas of use are primarily the bladder, liver, kidneys, and respiratory organs.

It boosts the immune system and its antioxidant effects are considered proven. There are hardly any scientifically recognized studies in Europe, however. The lignans (schisandrin, schisandrol, etc.) with their antioxidant effects are considered to be the most important group of active agents. With vitamins C, E, provitamin A (beta-carotene), niacin, riboflavin, and thiamine, the schisandra berry has a very high vitamin content. It also contains essential oils, flavonoids, pectin, resin, phytosteroles, tannins, and triterpene, as well as various minerals and trace elements. This wealth of active agents explains the schisandra berry's beneficial impact on the internal organs, emotions, and brain function, as well as the body as a whole.

The fruit of the five flavor berry hang together in a cluster, like currants. The range of flavors mean there are a great many ways of using them in the kitchen, and that they work well in numerous dishes. I like the berries both fresh and dried in muesli, yogurt, or fruit salad. They also make an outstanding berry sauce to accompany venison. Their spicy aroma makes them an interesting component in juices and fruit spreads. The schisandra berry is also known as the "Chinese magnolia vine" or "Chinese lime tree." The latter describes the aroma of the leaves when used to make teas and herbal preparations. Pour 1–2 teaspoons of boiling water over the crushed leaves and leave to infuse for 12 hours. Sweeten with sugar or honey, according to taste. This tea can also be used to flavor a variety of jams and juices.

This attractive plant has been available on the market as the schisandra berry or five flavor berry for some years now. As a perennial, winter hardy climbing plant, it needs a stable climbing frame—the schisandra berry can grow to over 16 feet (5 m) in height! The newer shoots become woody, such that the plant is able to support itself later on. It grows and overwinters best in the semi-shade in a light, moist soil, in a wind-protected location. You either need to keep an eye on its vigorous root suckers or use them for propagation. The schisandra berry is also suitable as greenery along fences, screens, and building facades. There are plants with male and female blossoms and so for a good fruit set you need to plant two specimens. Several female plants can be pollinated by one male plant. As of the second year after planting, the pale yellow, highly fragrant blossoms appear on the female plants between late spring and summer. In Europe, the fruit are harvested in September through October. In the berry's main growing area in northern China, the fruit is harvested only after the first frost.

Flavorsome cooking with schisandra berries

Mushroom soup with schisandra berries

Serves 4

1 piece fresh ginger (approx. 1 in / 2 cm)
1 garlic clove
1 stalk lemongrass
1 fresh red chili
14 oz (400 g) white fungus (tremella)
2 tbsp soy oil
3⅓ cups (800 ml) chicken stock
4–5 dried, pitted dates
3 tbsp dried schisandra berries
½ tsp mustard seeds
½ tsp cilantro seeds
1–2 tbsp lime juice
2–3 tbsp light soy sauce
2 tbsp rice vinegar
salt and freshly ground black pepper

■ Peel the ginger and garlic and dice finely. Wash the lemongrass, slice in half lengthwise, and crush.

■ Wash the chili, slice in half lengthwise, remove the seeds, and chop finely. Wash the mushrooms thoroughly, rub dry with a paper towel, and chop into smaller pieces if required.

■ Heat the oil in a saucepan and sauté the prepared ginger, garlic, chili, and lemongrass lightly. Add the stock, simmer for about 10 minutes, and pour through a sieve.

■ Cut the dates into smaller pieces if required. Add to the soup, together with the mushrooms and the schisandra berries. Bring to a boil and simmer over medium heat for a further 10 minutes.

■ Add the mustard and coriander seeds and then season with the lime juice, soy sauce, rice vinegar, salt, and pepper.

Variation

You could use currants or cranberries instead of schisandra berries to make this soup.

Fillet of venison with schisandra sauce

SERVES 4

2 sprigs thyme
2 sprigs rosemary
1¾ lb (800 g) oven-ready venison fillet
2 tsp ground venison spice mix
salt and freshly ground black pepper
2 tbsp clarified butter
⅔ cup (150 ml) venison stock
1¼ lb (600 g) Brussels sprouts
scant 2 cups (250 g) schisandra berries
⅓ cup (80 ml) ruby port
1–2 tbsp sugar
1–2 tsp cornstarch
1 tbsp butter
thyme for the garnish

■ Pre-heat the oven to 212 °F (100 °C). Wash the herbs and shake dry. Wash the meat, pat dry, and season with the venison spice mix, salt, and pepper.

■ Heat the clarified butter in a roasting pan and brown the meat on all sides. Add the herbs and pour over a little stock. Cook in the pre-heated oven for 40–50 minutes until pink.

■ Wash the Brussels sprouts. Cut off the stalk ends, remove the outer leaves, and slice the Brussels sprouts in half. Place them in boiling salt water and blanch for about 5 minutes. Drain through a colander, rinse under cold water, and leave in the colander to drain again.

■ Pick any leaves or stalks from the schisandra berries, wash the fruit in a sieve, and leave to drain. Bring to a boil in a pan with a little water, and simmer until the berries are soft. Pass through a strainer or food mill to remove the stalks and bitter seeds. Bring the fruit back to a gentle boil. Combine the cornstarch with the port and sugar and stir into the berry purée. Simmer for about 1 minute, and then set aside.

■ Remove the meat from the oven. Melt the butter in a skillet and brown the venison fillet for about 1 minute, repeatedly basting with the butter. Remove the meat from the skillet and set aside.

■ Sauté the Brussels sprouts briefly in the same skillet, and season with salt and pepper. Cut the venison fillet into slices and arrange on individual plates with the Brussels sprouts and schisandra berry sauce. Serve garnished with thyme. Boiled potatoes or potatoes au gratin are suitable accompaniments for this dish.

Kiwis—refreshing and invigorating

The first kiwifruit plants (genus *Actinidia*) made their way to New Zealand from the Far East at the beginning of the 20th century. In order to market them successfully, they were named after New Zealand's flightless, brown, national bird—the kiwi. The large-fruited kiwi (*Actinidia deliciosa*) with their furry, brown berries and green or yellow flesh are now very popular in Europe and the USA. Although the fruit comes from the tropics, you can grow this vigorously growing climbing plant easily in your own garden. Trained along espaliers, they can transform a bare corner into a shady paradise where you can enjoy the fruit straight from the plant.

All varieties of kiwi contain abundant Vitamin C, with just two of them meeting our daily requirement. They also contain Vitamin E, folic acid, B vitamins, minerals, and trace elements. Kiwis boost the circulation, metabolism, and the heart. Their high dietary fiber content has a detoxifying effect. Kiwis contain few calories (about 55 kcal per 3½ ounces / 100 g) and are therefore well-suited to low fat cooking. Their high pectin content lowers cholesterol and means that kiwis can be used as a gelling agent for all kinds of treats and soft setting fruit spreads. With their delicious flavor, they make great chutneys, sauces, and jams.

The refreshing, sweet-and-sour taste of this wonderful summer fruit is especially suited to chilled desserts or in ice cream or sorbets. When served with milk, quark, or cream, however, they become bitter in taste. This can be prevented by blanching them briefly. The healthier option is to combine them with milk products immediately before eating—or to eat them separately with muesli or similar breakfast cereals. A tip for tough meat: rub it with half a fresh kiwi a few minutes before using. This works to break down the tough connective tissue and tenderize the meat.

Outside of winegrowing areas, kiwis are often sensitive to late frosts and are only able to ripen in protected locations in the garden. Some varieties are dioecious, meaning that both genders need to be planted for a successful harvest. Much less complicated to grow are the smaller, very sweet kiwi berries (*Actinidia arguta*). They are more robust, ripen earlier, and are less sensitive to frost. Their skin can also be eaten too. They reach optimum ripeness for eating about four weeks after harvesting when stored in a cool place. They can be stored for several months in a cool, dry place (up to 39 °F / 4 °C).

Key to the successful growing of kiwifruit plants is a south-facing location as well as a deep, humus-rich soil with a good water supply and a thick mulch layer of acidic material (e.g. bark mulch). To keep the soil around the plant permanently acidic, mix it with coniferous wood shavings and oak bark. The optimum pH value is 4.5–5.5. It is advisable to plant kiwis in the spring so that they have time to develop proper root systems and are firmly rooted by the fall. This vigorously growing climbing plant is best pruned in the summer.
Even with the self-pollinating varieties, you will obtain a much more abundant harvest if you grow a male plant as well (for up to eight female plants). I personally prefer the red-skinned, small-fruited varieties. They have a more intense flavor and the plants create a pretty display in the garden.

Flavorsome cooking with kiwis

Kiwi cheesecake

■ For the base: combine all the ingredients, knead together quickly, and shape into a ball. Cover in plastic wrap and chill for 30 minutes in the refrigerator.

■ Pre-heat the oven to 350 °F (180 °C) (fan). Grease the springform pan with butter.

■ Roll about two-thirds of the base mix out on a floured surface and place in the pan. Use the rest to form the edge. Prick the base all over with a fork.

■ For the filling, separate the eggs and beat the egg whites until stiff.

■ Combine the cream cheese with the egg yolks, quark, sugar, lemon juice, and lemon zest. Sieve the baking powder together with the cornstarch over the mixture, and stir in well. Fold in the beaten egg white.

■ Spread the creamed mixture over the base and bake for about 1 hour. Cover with aluminum foil as soon as the surface starts to darken.

■ Let the cheesecake rest in the pan for a few minutes. Then remove and leave to cool completely on a wire rack.

■ Peel the kiwis, cut them into thin slices, and arrange decoratively on the cheesecake. Dust with confectioners' sugar before serving.

MAKES 1 SPRINGFORM PAN (11 IN / 28 CM DIAMETER)

For the base

1⅓ cups (200 g) all-purpose flour
3½ tbsp (50 g) sugar
7 tbsp (100 g) butter
1 egg
pinch of salt
butter for greasing the pan

For the filling

7 eggs
1¼ cups (300 g) heavy cream cheese
1½ cups (400 g) quark
scant 1 cup (200 g) sugar
2 tbsp lemon juice
1 tbsp grated zest from an unwaxed lemon
1 tsp baking powder
1 cup (130 g) cornstarch
5–6 kiwis
confectioners' sugar for dusting

Kiwi salad with avocado

Serves 4

8 kiwis
2 sprigs of fresh cilantro
1 ripe avocado
1 tbsp lemon juice
salt and freshly ground black pepper
3–4 tbsp olive oil
2–3 tbsp orange juice

■ Peel the kiwis and cut them into thick slices about ¼ inch (5 mm) thick. Wash the cilantro and pat dry. Remove the leaves from one sprig and keep the other sprig for the garnish.

■ Cut the avocado in half, removing the pit and the skin. Cut the flesh into slices and drizzle immediately with lemon juice.

■ Arrange the kiwi slices decoratively on a serving platter, place the avocado slices in the middle, and season with salt and pepper. Drizzle with olive oil and orange juice, and garnish with the cilantro (leaves and sprig).

Table grapes—a princely pleasure

Table and wine grapes are the berries of the common grape vine (*Vitis vinifera*). They are among the most frequently cultivated fruits worldwide. Their distribution in Europe largely goes back to the pleasure-loving princely houses of the past. Aristocratic ladies and gentlemen did not want the sometimes very sour fruit wines drunk by their subjects, preferring to refresh themselves with the sugary sweet flavor of ripe grapes. They had the most magnificent specimens planted along their castle walls. Many of the now ancient grapevines in the proximity of castles and palaces derive from this predilection. Having ripe grapes on the table was one way of impressing noble visitors, as was a good wine. And approval came from the highest authority—the drinking of wine is mentioned in the Bible, so neither the fruit nor the alcoholic beverage made from it could possibly be viewed as sinful.

Table grapes remain a true delight today, either in the hand, pressed into non-alcoholic juice, or fermented as wine. There are also a great many options for using the fruit in the modern kitchen. Grapes work well in all kinds of salads and harmonize especially well with nuts, pears, figs, or fennel. Their full-bodied flavor enriches red wine sauces to accompany poultry and venison. They are also popular in cold or hot soups, while I like them best as a foamy cream dessert. In cakes and bakes they are especially good in combination with almonds or other nuts. A

couple of skewers with grapes and a full-flavor cheese or smoked fish and you have a tasty meal for unexpected guests. And our kitchens just would not be the same without dried grapes as a snack and energy food—grandma's spice cake without raisins? Unthinkable!

The fresh and dried grapes available commercially have often been treated with pesticides, so I recommend organically grown fruit—and growing them yourself, of course. This should also protect you from the unpleasant cold sores that can result from eating conventionally grown grapes. When shopping for grapes, look for ripened fruit, as grapes do not continue to ripen once harvested. With white grape varieties, the yellower they are, the sweeter they will be.

The color intensity is also indicative of flavor with red grape varieties as well. Grapes have a natural protective film that keeps the fruit fresh for a while, so it is important to wash them thoroughly before eating.

Grapes contain abundant dextrose, fructose, and dietary fiber and are therefore a source of mental and physical energy. Their healthy flavonoids protect the heart, circulation, and immune system. Grapes stimulate the digestion and the elimination of toxins. In addition to Vitamin C, folic acid, and B vitamins, they also contain a wealth of minerals such as potassium, calcium, magnesium, and iron.

Fans of unusual flavors should look out for Muscat grapes. They are sweet, with their very own flavor. Their vines are real sun lovers; but all grape varieties need a sunny, protected location in the garden. Suitable sites would be against a house or garden wall, or a screen. This enables the gardener to keep everything under control, from pruning through ripening to the fall harvest. In the spring, you need to remove all the weak and frozen side shoots, limiting the plant to two or three main shoots. If they are to be trained along a wire frame, this is the time to position and fasten them. In the early summer, you need to feed the plants regularly with an organic berry fertilizer. As the grapes ripen, you can thin out some of the leaves hanging directly over the fruit: more sun produces more sugar and thereby more flavor. And the longer you leave the grapes on the vine, the sweeter and more full-bodied they will become.

PiWi varieties have been introduced in recent decades in order to combat downy mildew, these being fungus-resistant grape varieties. These vines require no—or only very little treatment—with pesticides.

In very dry summers, I put out a bird bath to distract the birds and wasps from the juicy fruit—they quench their thirst with water rather than the valuable grapes. It may still be necessary, though, to protect the fruit through to full ripeness with an approved, specialist net. All openings need to be closed so that no birds get tangled in the netting. Having a water barrel close by is recommended for achieving large fruit size and greater sweetness, as regular watering encourages and stabilizes the fruit formation. After all, it is fat, juicy grapes we want to harvest—not raisins!

Flavorsome cooking with grapes

Vanilla soup with white and red grapes

SERVES 4

9 oz (250 g) seedless white grapes
9 oz (250 g) seedless red grapes
1½ quarts (1.5 liters) milk
1 vanilla bean
4 eggs
7 tbsp (100 g) sugar
5 tsp cornstarch
ground cinnamon for dusting

■ Wash the white and red grapes and slice them in half.

■ For the vanilla soup: bring the milk to a boil in a pan, together with the scraped out vanilla bean and its seeds.

■ Separate the eggs. Beat the egg whites in a bowl until stiff and then chill.

■ In a separate bowl, beat the egg yolks together with the sugar until foamy and then fold in the cornstarch.

■ Slowly stir in the boiling milk. Return the milk mixture to the pan and re-heat, stirring all the time. Do not allow it to boil.

■ Take the vanilla soup off the heat, remove the vanilla bean, and fold the beaten egg whites into the still hot soup. Serve in soup bowls with the grapes and a dusting of ground cinnamon.

Grape balls with blue cheese and nuts

MAKES 16

16 seedless white grapes
scant 1 cup (100 g) crumbled blue cheese
6 tbsp (100 g) cream cheese
3–4 tbsp chopped, blanched almonds
3–4 tbsp chopped pistachios

■ Wash the grapes, remove them from their stalks, and pat dry.

■ Place the blue cheese in a bowl together with the cream cheese, mash with a fork, and mix well together. Coat the grapes in this mixture.

■ Lightly toast the almonds in a skillet and set aside to cool. Then lightly toast the pistachios and set them aside to cool.

■ Roll half of the grape balls in the almonds and the other half in the pistachios. Arrange on plates and serve.

Berries from the wild

Sea buckthorn—a bountiful harvest

Heavily laden with orange berries among silvery leaves, the sea buckthorn (*Hippophae rhamnoides*) is visible from afar. Widespread along the North and Baltic Sea coastlines of Europe where it flourishes on the dunes, sea buckthorn also occurs sporadically inland in central and western Europe. It is able to develop roots and runners even in shingle and gravel and can resist stormy weather. This undemanding bush also forms a symbiotic underground relationship with various fungi and bacteria, which enables it to self-fertilize. This in turn enables it to colonize even bare, stony slopes along river valleys, where competition from shady trees is minimal. Sea buckthorn is often used to stabilize ground conditions, e.g. along the banks of highways, as well as to brighten such manmade areas with natural color.

The branches of the sea buckthorn bear sharp thorns, so picking the berries with bare hands is not advisable. As the berries are also pressure-sensitive, bursting easily, I cut off entire shoots and place them in the freezer. After a few hours, when the berries have frozen solid, I beat the branches against the kitchen counter until all the berries and leaves have fallen off. After sorting, the yield is typically 9–13 pounds (4–6 kg) of fruit per bush. It is best to process the berries immediately. Cutting back the fruit-bearing shoots every two years does not damage the bush at all; rather, it encourages the growth of new shoots. Timing is key if you are to benefit from as many of the berries' valuable ingredients as

possible. Sample the berries regularly until they have reached a good balance between acid and sweet.

Sea buckthorn berries have a fine, fruity aroma. Sprinkled with sugar, you can eat them raw—in fruit salad, for example, or with

quark, yogurt, and other dairy products. If you want to preserve the fruit as juice, purée, or fruit spread, you will need a high-performance food processor that is also able to grind the seeds. This will ensure that you enjoy all the vital nutrients, which are abundant in it: a great deal of vitamins C and E, B vitamins including B12, folic acid, and beta-carotene, as well as numerous minerals, trace elements, and valuable oil. Sea buckthorn's range of applications as a medicinal plant is correspondingly broad. It boosts the circulation and immune system and helps in the treatment of exhaustion, colds, and stomach conditions. The oil from the fruit flesh is used for skincare purposes and for a variety of skin conditions such as neurodermatitis. It is said to have a rejuvenating effect on the skin through its cell-renewing properties.

The unique combination of active ingredients and straightforward growing requirements makes the sea buckthorn popular with horticulturalists across the globe, in cooler regions in particular. Not everyone has access to wild sea buckthorn, which is why I recommend the plant to anyone with a spare sunny spot in their garden. Actually, two spots, to be precise—the sea buckthorn has both male and female types. You will need to position the plants close to one another, so that the wind can carry the pollen from the former to the stigmas of the latter. It is also important that both specimens blossom at the same time (see

Recommended varieties, p. 181). It is not necessary for every female plant to have its own male—one male plant is sufficient for five through ten females. Sea buckthorn has a tendency to proliferate, so to keep growth in check it is advisable to remove offshoots early on, with the sharp edge of a spade. A root barrier will prevent the shallow runners from proliferating. Sea buckthorn does not need any fertilizing; it flourishes even on poor, sandy soils. Nor does it require watering—it is the perfect bush for the "lazy bed" method of gardening. It is an attractive and thoroughly useful component in a wild fruit hedge, but should be kept away from seating areas and children's playgrounds on account of its long, sharp thorns. It is better suited to borders and parts of the garden where you want to discourage access.

Sea buckthorn leather

MAKES 1 TRAY

generous 1 lb (500 g) sea buckthorn berries
1 lb 10 oz (700 g) sugar
1 oz (30 g) pectin
sugar for rolling

This confectionery can be kept for longer periods in an airtight container. Layer them on sheets of wax paper to prevent the pieces from sticking to each other.

■ Pick any stalks and leaves from the sea buckthorn berries, wash the fruit thoroughly, and pat dry. Place in a bowl, mix with 1½ pounds (650 g) of the sugar, and leave to infuse overnight.

■ The next day, place the berry mixture in a saucepan, pour in just enough water to cover, and heat until the berries burst. Purée the berry mixture in a blender and then pass through a food mill or muslin cloth.

■ Combine the rest of the sugar (2 ounces / 50 g) with the pectin, add to the fruit purée, and bring to a boil. To test whether the mixture has reached the right consistency, place a spoonful of the fruit purée in a bowl of cold water. Then shape it into a small ball. The mixture is ready when it yields to the touch and is easy to shape. If it falls apart, the mixture will need to be boiled for longer.

■ When ready, spread the mixture about ½ inch (1 cm) thick on a tray lined with wax paper and leave to dry in a warm, airy room for several days.

■ Cut into little rectangles or cubes and then roll them in sugar.

Variation

Instead of rolling the pieces in sugar, you could also coat them in melted couverture chocolate.

Flavorsome cooking with sea buckthorn

Apple and sea buckthorn cupcakes

MAKES 4 MUFFIN MOLDS (⅔ CUP / 150 ML EACH)

butter and sugar for the muffin molds
1 egg
7 tbsp (100 g) brown sugar
3½ tbsp (50 ml) vegetable oil
¾ cup (175 ml) milk
1¾ cups (250 g) all-purpose flour
1 tsp baking powder
generous ½ cup (50 g) ground, blanched almonds
1 tbsp vanilla sugar

2 ripe apples
1¼ cups (300 ml) sea buckthorn juice
6 tbsp (100 ml) whipping cream

■ Pre-heat the oven to 350 °F (180 °C) (fan). Grease 4 individual muffin molds with butter and sprinkle with sugar.

■ Beat the egg with the brown sugar, vegetable oil, and milk. Combine the flour with the baking powder and sieve over the egg mixture. Add the almonds and the vanilla sugar, and mix together well.

■ Divide the mixture between the muffin molds and bake for about 25 minutes, testing with a cocktail pick to see if it is cooked.

■ In the meantime, wash and peel the apples, cut them into quarters, and remove the cores. Dice the flesh finely. Heat the sea buckthorn juice with the diced apple in a pan.

■ Allow the baked cupcakes to cool slightly. Remove from the molds and place on individual plates. Pour the apple-sea buckthorn sauce over them. Beat the cream until stiff, spoon a little onto each of the cupcakes, and serve.

Quark strudel with sea buckthorn

MAKES 1 STRUDEL

For the pastry
generous 1⅓ cups (200 g) all-purpose flour
pinch of salt
2 tbsp corn oil
7 tbsp (100 ml) warm water

For the filling
2 cups (500 g) quark
⅔ cup (150 g) sugar
7 tbsp (100 ml) sea buckthorn juice
zest of 1 unwaxed orange
2 eggs
2 tbsp cornstarch
pinch of salt
3½ tbsp (50 g) butter
generous ½ cup (50 g) breadcrumbs
flour and confectioners' sugar for dusting

■ For the pastry, combine the flour and the salt in a mixing bowl. Add the corn oil and water and work, for at least 5 minutes, into an elastic, non-sticky dough. Shape into a ball, cover in plastic wrap, and leave to rest at room temperature for about 30 minutes.

■ For the filling, place the well-drained quark in a bowl and fold in the sugar, sea buckthorn juice, and orange zest.

■ Separate the eggs and stir the egg yolks into the quark mixture together with the cornstarch. Beat the egg whites with the salt until stiff and then carefully fold into the quark mixture.

■ Heat 2 tablespoons (30 g) of butter in a skillet and lightly brown the breadcrumbs, stirring all the time.

■ Spread out a large, clean dish towel and dust with flour. Place the pastry dough on top, roll out, and then stretch it out over the backs of your hands as thinly as possible.

■ Cut off the thicker edges of the dough. Sprinkle the browned breadcrumbs over it. Pre-heat the oven to 350 °F (180 °C) (fan). Line a baking sheet with wax paper.

■ Spread the quark and sea buckthorn mixture over the dough. Fold in the sides just enough to reach the filling and use the dish towel to carefully roll into a strudel. Roll onto the baking sheet with the join underneath.

■ Melt the rest of the butter and use to brush the strudel. Bake in the center of the pre-heated oven for about 45 minutes until golden brown. Leave the strudel to cool and served dustec with confectioners' sugar.

Barberries–tart and healthy

The barberry (*Berberis vulgaris*) grows throughout New England and Europe, even at altitudes above 8300 feet (2500 m). It is often planted in urban parks and gardens. It was only during my training as a horticulturalist in Germany's Eifel region, however, that I got to know and appreciate its fruit. There the branches were so laden that the red, ovoid berries were hanging temptingly close to my mouth—it was only the many thorns that stopped me from plucking them with my teeth. Carefully pick one of the berries and your tastebuds will soon tell you why this fruit is also known as the "sour berry" in some places—barberries are extremely tart, although this can be very refreshing in the heat of summer! With these wild berries, too, it is the processing that enables you to appreciate their intense flavor.

Harvesting is easiest when you use garden shears and thick leather gardening gloves to cut off whole branches. Freeze these and then shake the frozen berries off—over a blanket spread over the ground, for example. Then you just need to pick any leaves or stalks from the berries. They are not quite as sour after freezing and are easier to process. The white berries are naturally less sour.

Barberry seeds are considered to be slightly toxic and so should be removed before eating. This is best done with a steam juicer. You can also cook the washed berries in a little water for about 15 minutes until soft and then pass them through a muslin cloth or food mill. The resultant fruit purée can be used in a great many ways. It can also be used as a substitute for lemon juice and is similarly rich in Vitamin C. Boiled with sugar and sealed in airtight containers, it will keep for longer. You can combine it with other, milder, fruit varieties as a source of flavor and tartness. A juice made from the purée is especially refreshing and can also be used to make vinegar and mead. In folk medicine, barberry juice has been used for centuries as a coolant against fever as well as for stomach and digestive complaints, and liver and gall conditions, among others. As well as Vitamin C, the berries contain carotenoids, fruit acids, pectin, and minerals.

Older bushes become naturally seedless, at which point the fruit can be eaten whole. You will find them dried in Asian shops where they are sold for use in numerous traditional dishes such as "Zereshk Polo," the famous Persian rice dish. Several thousand

tons of barberries are harvested annually in the Middle East. The fruit is also used to flavor sweet-and-sour fish dishes and roasts. In France, the seedless "Asperma" variety is used to make the famous candy "Confitüre d'épine-vinette."

A more compact sister variety has also been available on the market as a wild fruit for several years now—the Korean barberry (*Berberis koreana*). The "Azisa" and "Rubin" varieties are especially high in Vitamin C and fruit acid (up to 11 percent of the fruit's weight). They retain their berries longer than the indigenous barberries and can also still be harvested in the winter. Barberries are generally undemanding and even grow in very heavy soils. Many barberry varieties are poisonous, so you should source your bushes from a good nursery. The barberry is also known as the berberis or berberry. You can recognize the wild barberry by the bright yellow color of its wood, which sets it apart from other thorny fruit bushes.

Flavorsome cooking with barberries

Barberry muffins

Makes
12 muffins

¾ cup (175 g) soft butter
²/₃ cup (150 g) sugar
4 eggs
zest and juice of 1 untreated lime
1¼ cups (175 g) all-purpose flour
7 tbsp (40 g) ground almonds
½ tsp baking powder
6 tbsp (50 g) dried barberries
a little soft butter and flour for the muffin pan

■ Pre-heat the oven to 350 °F (180 °C) (fan). Beat the butter with the sugar until creamy.

■ Whisk the eggs together with the lime juice and zest. Fold into the butter mixture together with the flour, almonds, and baking powder. Mix well and then fold in the barberries.

■ Grease the muffin pan and dust with flour. Place the dough in the muffin pan and bake in the oven for about 25 minutes until golden brown.

■ Leave to cool in the pan before serving.

Variation

You could also use cranberries, elderberries, or serviceberries instead of barberries.
Adding half a teaspoon of bicarbonate of soda will make the muffins especially light.
Other tasty additions include 5½ oz (150 g) roughly chopped chocolate. If you prefer a nutty flavor, you could use ground hazelnuts instead of almonds.

Chicken and barberry salad

Serves 4

4 sweet oranges
1 tsp coriander seeds
½ tsp chili powder
2 tsp honey
3–4 tbsp vegetable oil
2 tbsp dried barberries
4 chicken scallops, 4 oz (120 g) each
salt
2 tbsp pumpkin seeds
3 oz (80 g) pitted black olives
1 red onion
1–2 tbsp cider vinegar

■ Peel the oranges with a sharp knife and cut into sements. Squeeze the juice into a bowl and discard the pith.

■ Crush the coriander seeds with a pestle and mortar and combine with the chili powder, 1 tablespoon of the freshly squeezed orange juice, 1 teaspoon of honey, and 1 tablespoon of vegetable oil.

■ Place the barberries in the rest of the orange juice to infuse.

■ Rinse the chicken scallops and pat dry. Brush with the spicy sauce, season with salt, and place in a steamer. Cover and steam for about 10 minutes.

■ Chop the pumpkin seeds and drain the olives well. Peel the onion, cut in half, and then slice into thin strips. Leave the chicken to cool and then cut into bite-size pieces.

■ For the dressing, combine the onion strips, the rest of the oil, the remaining honey, and the cider vinegar, and mix into the barberries, seasoning with salt.

■ Combine all of the prepared salad ingredients and arrange on individual plates. Drizzle with the dressing and serve.

Mahonia—the Native American berry

Lewis and Clark, the leaders of the expedition into the wilds of Oregon, were not at first aware of what wonderful blue berries they had been given for their journey by the Shoshone people. On their return to civilization (1806) it was these berries that had ensured their survival. Full of gratitude, they took a bush to the American botanist McMahon (also known as M'Mahon), for whom the plant was later named. The mahonia crossed the ocean in 1825 and made its entry into Europe via the formal, artistically designed "villa gardens" of the day—from where they became widespread in parks and gardens.

The decorative, evergreen mahonia (*Mahonia aquifolium*) loses very few leaves and also turns red in parts in the fall, making it a favorite among landscape gardeners. Due to its prickly leaves, however, it should not be planted near seating areas. In the spring, the frost hardy mahonia is a delight to look at with its sulfur yellow fronds of blossoms, visible from afar and supplying honeybees and bumblebees with their first nectar. In its American homeland it is the state flower of Oregon where it is also known as the Oregon grape. Unlike the poisonous holly (*Ilex*), for which it is sometimes mistaken, the wood of the mahonia bush is deep yellow in color, which becomes all the more evident when cut.

The blue mahonia berries contain an intensely red juice that can be used to color berry mixtures and all kinds of foods. Mahonia used to be a popular plant in winegrowing areas; just a little of its juice added to the wine was enough to turn a pallid, insipid pressing into a robust red wine with a distinct "berry nose." The juice is full of Vitamin C and is tart to the taste. This makes it a versatile alternative to lemon juice, similar to barberries. It can be used to give all kinds of fruit juices a refreshing touch, as in an apple spritzer, for example. For jellies and compotes the sour berries are mixed with other fruit, harmonizing especially well with pears or quinces. I prefer the boiled, sugary juice best as a fruit sorbet.

Gloves and gardening shears are advisable for harvesting because the ripe blue berries are sensitive to pressure and they do stain. I cut off the berry clusters whole and place them carefully in a bucket. I then transfer them to the freezer for a few hours. When frozen, the berries come away from their stems easily and cleanly. Boil them in a little water and pass them through a muslin cloth or food mill. This sieves out the seeds, which are slightly toxic (berberine) in large quantities. A steam juicer is ideal for this as well. This "lemon juice" makes a healthy and robust basis for all kinds of sauces, fruit spreads, and desserts. To preserve it, boil the juice again with sugar and store in airtight containers. From a nature conservation perspective, I recommend the harvesting of wild mahonia berries in particular, as this stops the bushes from spreading in the natural environment—mahonia does not belong in the forest.

The garden is the best place for these decorative, undemanding berry bushes. For the fruit to be able to ripen properly, the bush does need a slow-release, organic fertilizer. Without this the leaves can grow too quickly, making the plant susceptible to powdery mildew. Regular watering in the summer ensures a plump berry size and therefore a good juice harvest.

Flavorsome cooking with mahonia berries

Iced mahonia mousse

MAKES 1 LOAF PAN (10 × 4 IN / 26 × 10 CM)

3½ oz (100 g) mahonia berries
7 oz (200 g) raspberries
scant 1 cup (100 g) confectioners' sugar, or to taste
4 gelatin leaves
1 vanilla bean
generous ¾ cup (200 g) quark
generous ¾ cup (200 g) yogurt
1 tbsp lime juice
7 tbsp (100 g) sugar
4 egg whites
generous ¾ cup (200 g) whipping cream

■ Wash the mahonia berries and boil with a little water for a few minutes until soft. Then pass through a food mill and set the purée aside.

■ Pick any leaves or stalks from the raspberries, wash the fruit if required, and pat dry. Warm the raspberries briefly in a pan, leave to cool slightly, and then purée them in a blender. Return them to the pan, stir in the mahonia purée and the confectioners' sugar, and re-heat briefly. Then leave to cool completely.

■ Soak the gelatin in cold water. Slit open the vanilla bean lengthwise and scrape out the seeds. Mix the quark together with the yogurt, lime juice, vanilla seeds, and sugar.

■ Dissolve the soaked gelatin in a pan over low heat and transfer to a large mixing bowl. Stir in 2–3 tablespoons of the quark mixture before folding in the rest of it.

■ Beat the egg whites and also the cream, separately, until stiff. Carefully fold both into the quark and gelatin mixture. Place in a loaf pan and stir in the raspberry and mahonia berry purée in swirls with a fork.

■ Place in the refrigerator for about 2 hours to set and then in the freezer for about 30–45 minutes. The mousse should not be allowed to freeze solid.

Ricotta cake with berry sauce

Makes 1 springform pan (10 in / 26 cm diameter)

For the base
1¼ cups (180 g) all-purpose flour
2 tbsp cocoa powder
3½ tbsp (50 g) sugar
7 tbsp (100 g) butter
1–2 tbsp light cream

For the topping
4 eggs
pinch of salt
scant 1 cup (200 g) sugar
2¼ lb (1 kg) ricotta
2 tbsp lemon juice
2 tbsp cornstarch

For the mahonia berry sauce
3½ oz (100 g) mahonia berries
7 oz (200 g) raspberries
6 tbsp sugar, or to taste
4 tsp (20 ml) raspberry Schnapps
1 tsp cornstarch
1 tbsp vanilla sugar

■ For the base: combine the flour with the cocoa and sugar. Add the butter in pieces, together with the cream, and quickly work into dough. Add a little cold water or flour as required. Cover in plastic wrap and chill for about 30 minutes.

■ Pre-heat the oven to 350 °F (180 °C). Line a springform pan with wax paper and roll out the dough to the size of the pan. Place on the base of the pan and prick all over with a fork.

■ For the topping: separate the eggs. Beat the egg whites with the salt until stiff. Beat the egg yolks with the sugar until foamy and then fold in the ricotta and lemon juice. Fold in the cornstarch and then the beaten egg whites.

■ Spread the mixture over the base and smooth the surface. Bake in the oven for about 1 hour. If necessary, cover the cake with wax paper to stop it turning too dark. Leave to cool.

■ For the sauce: cook the washed mahonia berries with a little water until they are soft. Then pass them through a food mill or muslin cloth.

■ Pick any leaves or stalks from the raspberries, wash the fruit, and heat gently in a pan. Crush slightly with a potato masher, then stir in the mahonia purée, sugar, and raspberry Schnapps, and bring to a boil.

■ Combine the cornstarch with the vanilla sugar and mix with a little cold water. Use to bind the sauce, simmering gently for about 1 minute, stirring frequently. Then leave to cool. Remove the cake from the pan, cut into slices, and serve drizzled with the sauce.

Elderberries—trendy berries

The elder is enjoying something of a revival—its abundant flower sprays flavor a range of summer drinks. The fruit of the elder, the deep-blue through to violet-black elderberry, is also nutrient rich, being packed full of healthy ingredients.

The black elder (*Sambucus nigra*) has been known to mankind since primeval times. A common feature at the edges of woodland, it soon found its way into nearby villages; well fertilized fields and dung heaps have proved to be an ideal bed for its seeds to this day. Reaching a height of up to 33 feet (10 m) it was once the impressive focal point of the traditional country garden. Its gifts were so diverse, beneficial, and abundant that, in German, this venerated medicinal and protective plant ("Holunder") is named for the Earth Mother of the Germanic tradition, Frau Holle.

The black elder has been used as a medicinal plant since the Stone Age. Hippocrates referred to it as the "medicine chest." In the winter, and especially during the season for colds and flu, the hot berry juice is a miracle weapon against infections. Enjoyed with a little honey, it soon makes the patient feel more comfortable. The berries have a fever-reducing and pain-killing effect. Containing Vitamin C, potassium, B vitamins, folic acid, fruit acid, essential oils, anthocyanins, further flavonoids, and more especially the red-violet coloring agent sambucyanin, elderberries can increase resistance to infections, flu viruses, and allergies. The flowers are also used as a tea for head colds. They are sudorific

(sweat inducing), anti-inflammatory, and function as an expectorant. In the kitchen, elderberries can be used in a multitude of ways—in rich, red fruit jams and juices, fruit wine, punch, soups, jellies, fillings, or an ingredient in all manner of desserts. Combined with apples, pears, and quinces they produce delicious creations. When harvesting, you should only gather fully ripe berries, as these have the most nutrients, the best taste, and are more easily digested. They do stain, so pick them as whole sprays and protect yourself in the kitchen with gloves and an apron.

Elderflowers are a popular addition to lemonade, syrup, punch, and cocktails. Used in the past to flavor mead, a wide variety of drinks and desserts can be flavored with these pretty, creamy white blossoms. To do this, you simply infuse them in the liquid in question for a few hours before heating slightly: for example, in an elderflower and vanilla pudding. Deep-frying the blossom sprays in pancake batter to make elderflower fritters is an especially delicious way of preparing them.

Caution! The berries of the black elder (*Sambucus canadensis*) contain a poisonous alkaloid—even when fully ripe, they should never be eaten raw, as this can lead to nausea and vomiting. Unripe berries should never be consumed, whether raw or cooked. You should only use fully ripe berries and cook them thoroughly before eating, to ensure that the toxins are fully destroyed.

The red elderberry (*Sambucus racemosa*) is also edible. Cooking does not eliminate the toxins in its seeds, however, so the seeds must be removed before eating—by using a steam juicer, for example. The berries are somewhat stronger in taste and are therefore ideal for sauces to accompany venison dishes. There is only one type of elder indigenous to Europe that is poisonous in any form: the European Dwarf Elder (*Sambucus ebulus*). It grows as a non-woody shrub. The upright bush reaches no more than 40 inches (1 m) in height. It is distinguished by its upward-facing flower clusters, which later carry the fruit skyward. All the other indigenous varieties have hanging flower clusters and fruit.

Despite all its positive properties, the black elder largely disappeared from towns and gardens during the 20th century. As a result of mechanization and modernization, many indigenous species had to make way for on-trend, foreign, or "tidy" shrubs. Naturally sown plants were removed and even ancient elderberries drastically cut back. With today's return to natural looking gardens and landscapes, however, the elder is also making a comeback. The plant is extremely low maintenance—a shady spot rich in nutrients is enough for the berries to form in abundance. Watering in summer is beneficial, as are organic fertilizers or horn shavings for nitrogen.

Recommended varieties: see p. 180.

Elderberry soup

Serves 4

14 oz (400 g) elderberries
1 2/3 cups (400 ml) unsweetened blackcurrant juice
generous 3/4 cup (200 ml) unsweetened apple juice
2/3 cup (150 ml) water
2 pears
juice of 1 unwaxed lemon
2–4 tbsp honey
1 vanilla bean

■ Pick any leaves etc. from the elderberries then wash the fruit and drain well. Remove the elderberries from their stalks and bring to a boil in a pan with the blackcurrant juice, apple juice, and water.

■ Simmer gently for a few minutes, or until the berries burst. Then pass through a food mill. Bring the resultant purée back to a boil.

■ Peel the pears, remove the cores, and cut into pieces. Slit open the vanilla bean lengthwise and scrape out the seeds. Then add the vanilla bean and seeds to the elderberry purée, together with the lemon juice, 2–4 tbsp honey (according to taste), and pears.

■ Simmer over medium heat for a few minutes, stirring frequently. Then remove the vanilla bean, flavor the soup with honey to taste, and serve warm or cold.

Flavorsome cooking with elderberries

Elderberry pudding

Serves 4

butter for greasing the dish
4 eggs
3½ tbsp (50 g) sugar
2 tbsp vanilla sugar
½ cup (75 g) all-purpose flour
pinch of salt
pinch of baking powder
6 tbsp (100 g) quark
⅓ cup (75 ml) milk
1½ lb (650–700 g) elderberries
a little flour and confectioners' sugar for dusting

■ Pre-heat the oven to 350 °F (180 °C) (fan). Grease a shallow baking dish.

■ Beat the eggs together with the sugar and the vanilla sugar until foamy. Add the flour, salt, baking powder, quark, and milk and mix all the ingredients together well.

■ Carefully rinse the elderberries, strip them from their stalks, and drain well. Dust with a little flour and fold into the pudding mixture.

■ Pour the mixture into the greased baking dish and cook in the pre-heated oven for about 35 minutes. Served dusted with confectioners' sugar.

Steak with pears and elderberry sauce

SERVES 4

11 oz (300 g) ripe elderberries
1 shallot
3 lemons
4 pears
4 cups (1 liter) water
2¼ cups (500 g) sugar
1 cup (250 ml) Marsala or dry sherry
4 x 5 oz (150 g) beef steaks
salt, pepper, thyme
1–2 tbsp rapeseed oil
1 tsp butter
5 tbsp red wine
1 tbsp balsamic vinegar
1 tsp honey
4–6 sprigs of thyme for the garnish

■ Pre-heat the oven to 212 °F (100 °C) (fan). Wash the elderberries, pat them dry, and strip from their stalks. Peel and finely chop the shallot. Squeeze the lemons.

■ Wash and peel the pears (leaving the stalks on), and drizzle with the juice of one lemon.

■ Place the water in a pan together with the sugar, Marsala, and the rest of the lemon juice, and bring to a boil. Place the pears in the liquid as soon as it starts to boil and simmer gently for 10–15 minutes.

■ Reduce the oven temperature to 176 °F (80 °C) (fan). Carefully remove the pears from the pan with a slotted spoon and leave to drain. Transfer them to a shallow baking dish and keep warm in the oven. Set the cooking liquid aside.

■ Rinse the steaks, pat dry, and season with salt, pepper, and thyme. Heat some of the oil in a skillet and brown each one quickly for 2–4 minutes (according to taste). Remove from the skillet, wrap in aluminum foil, and leave to rest in the oven for about 10 minutes.

■ In the meantime, sauté the finely chopped shallot with the butter in the cooking juices. Add the elderberries and deglaze with the red wine plus ⅔ cup (150 ml) of the pear cooking liquid. Simmer for about 10 minutes.

■ Pass through a sieve and season with balsamic vinegar and honey. Take the steaks out of the oven, remove from the foil, and add any meat juices to the elderberry sauce.

■ Arrange the steaks and pears on individual plates and surround with the elderberry sauce. Sprinkle with freshly ground black pepper and garnish with sprigs of thyme.

Juniper—a conifer with berries

Although, botanically speaking, the common juniper (*Juniperus communis*) bears "pseudo-fruit" or "cones," its fruit is generally included with berries and is processed in much the same way.

A juniper bush needs lots of light and wide open spaces, its favorite habitat being grazed grassland. Widespread livestock farming in the past meant that the juniper was able to spread from its original mountainous home to more diverse terrain—in places, its conical shape characterizes entire landscapes in central Europe. For various reasons, the juniper berry has become rarer today and is now protected. Larger populations are now to be found in protected areas only, with the gathering of wild berries being generally restricted. As such, it is best to plant this ancient shrub, valuable for both culinary and medicinal purposes, in your own yard or garden.

Juniper wood and berries were in use as incense long before frankincense and myrrh found their way from the Orient to the West. Farm buildings needed to be protected from evil in general and disease in particular. In the European Alps, glowing juniper twigs are still taken through farmyards and stables at Christmas time to bless the buildings and the livestock housed within them. Solitary juniper bushes alongside roads or at road junctions are said to house gnomes and elves, and these plants are always protected from construction projects.

The medicinal properties of the juniper berry have been known to mankind for thousands of years. Today, it is used mainly in the treatment of rheumatic conditions, digestive complaints, and to boost the metabolism. It improves circulation; it is warming; and it can act variously as a muscle relaxant, diuretic, and detoxifier. A juniper bath can help with back pain. The juniper berry also provides an energy boost, particularly for those suffering debilitating tiredness. It gets us back on our feet after illness. Juniper Schnapps forms the basis for the renowned Swedish bitters. It should be noted, however, that the juniper berry contains substances that irritate the kidneys and so should not be consumed in large quantities over long periods of time. It should be avoided altogether in the case of kidney disease and pregnancy.

In the kitchen, juniper berries are used mainly to flavor savory dishes such as game, marinated pot roasts, and sauerkraut. In the past juniper was also important for preserving meat. My grandmother in Germany's Eifel region used to smoke her own hams with juniper berries and twigs she had gathered herself. She often used to wonder which "two-legged little mouse" had been nibbling at the ham! Today I prefer the berries as a fruit spread or "Latwerg," which I encountered in Switzerland. To make the latter, the ripe berries are ground and then mixed with caramel sugar, raw sugar, glucose syrup, and water to form a juice. This is then boiled until it acquires a spreading consistency. "Latwerg" is especially delicious served with a strong alpine or goats' milk cheese. The spicy juniper berries are also popular in alcoholic form. The gin once so beloved of the Queen Mother, the Dutch "Genever,"and the German "Eifelgeist" all derive from the juniper bush.

With its modest requirements, the juniper bush needs no more than a sunny spot in the garden where it can grow in peace. It takes three years for the blossoms to turn into berries. There is a wide range of shapes and colors available commercially, from the broad mountainous variety to the slim, conical shape. Juniper is dioecious and so you will need to grow both genders for a good fruit set. It is best to seek out a specialist nursery if you want to be sure of the plant's suitability for culinary use. There you will be able to clearly distinguish the edible juniper from the poisonous Savin juniper (*Juniperus sabina*) and the Chinese juniper (*J. chinensis*).

Flavorsome cooking with juniper berries

Brussels sprouts with bacon and juniper

SERVES 4

1¾ lb (800 g) Brussels sprouts
salt
2 shallots
5½ oz (150 g) diced bacon
2 tbsp rapeseed oil
1 tbsp juniper berries
1 cup (250 g) cream
freshly ground black pepper

■ Wash the Brussels sprouts, remove their stalk ends and outer leaves, and cut in half. Blanch in boiling salt water for about 10 minutes until firm to the bite, rinse under running cold water, and leave to drain.

■ Peel the shallots, dice finely, and sauté in hot oil together with the diced bacon for 2–3 minutes. Stir in the Brussels sprouts and sauté briefly.

■ Grind the juniper berries with a pestle and mortar and add to the Brussels sprouts, together with the cream. Simmer for about 1–2 minutes, season with salt and pepper, and serve.

Tomato and bell pepper soup with juniper cream

SERVES 4

3 red bell peppers
1 onion
1 clove of garlic
2 tbsp olive oil
14 oz (400 g) tomato purée
2½ cups (600 ml) tomato juice
7 tbsp (100 ml) dry white wine
2 bay leaves
1 tbsp ground paprika
1 tsp sugar
salt, pepper, nutmeg
6 tbsp (100 g) heavy whipping cream
1 tbsp juniper berries

■ Wash the bell peppers and chop them roughly, removing the pith and seeds. Peel the onion and garlic, chop finely, and sauté in hot olive oil. Add the bell pepper pieces and sauté briefly.

■ Add the tomato purée, tomato juice, and the wine. Add the bay leaves and simmer for about 10 minutes. Season with paprika, sugar, salt, pepper, and grated nutmeg.

■ Beat the cream until slightly stiff. Grind the juniper berries with a pestle and mortar and fold into the cream. Serve the soup in individual bowls, garnished with a swirl of the juniper cream.

Hawthorn—the heart berry

There it is, in the middle of the wild hedge—the hawthorn (*Crataegus* sp.). Its lighter wood makes it easily distinguishable from the darker blackthorn, or sloe. Together with other bushes like blackberries, buckthorn, and the dog rose, these plants form thorny, impenetrable hedges. Such types of natural barrier were used to protect human settlements and today such wild hedgerows provide vital shelter for many animal species.

As soon as the spring comes, however, the tough hawthorn bush is transformed; the fresh green shoots with their fragrant white blossoms are a gentle and refreshing sight. The hawthorn is the most important plant-based therapeutic agent for the heart and has long been used as such in traditional medicine, with no known side effects to date.

But the hawthorn also has a long tradition in the kitchen. The dried berries, full of vitamins, used to be finely ground in the winter and used to bake healthy, flavorful flatbreads and fruit loaves. Hawthorn berries were an important source of energy in the winter for the Native Americans too. They contain, for example, flavonoids, essential oils, amines, tannins, minerals, pectin, organic acids, and Vitamin C. To benefit from their medicinal effects, however, preparations and infusions made from the leaves, flowers, and fruit need to be taken over a longer period of time.

A fruit purée made from the boiled, sieved fruit can be used in many ways (see recipe on p. 148). It flavors mixed fruit jams and ensures a good set. It can be used to refine compote, red fruit jellies, and other desserts. Yellow-fruited varieties can also have visual appeal.

The hawthorn's general revitalizing and stress-reducing effects can also be enjoyed as a liqueur. Combine a few handfuls of ripe berries with sugar and vodka, for example, together with suitable flavorings like lemon, vanilla, cinnamon, or lemon balm. Pour the mixture into an airtight container and shake occasionally. The liqueur will be ready to enjoy after about two months—and it will get more than just your heart going. The boost to the whole circulatory system brings more oxygen into the body and reduces blood pressure. The accumulation of cholesterol is impeded and collagen in the tissues is protected from damage.

The hawthorn blossoms are also edible. They can be used to flavor drinks and sweet dishes by submerging them in the liquid in question and bringing it to a boil. Definitely worth a try for hawthorn blossom jellies, syrups, and liqueurs as well!

The berries are harvested from September through October. If you have picked them too early and the unripe berries are sour and astringent, you can freeze them for a couple of days. This breaks down the tannins and the sugar comes to the fore. For home growing I recommend seedlings sourced locally as these generally develop best. The variety with red thorns known as "Paul's Scarlet" with its full, pink-red flowers can be seen in many front gardens, but it generally bears no fruit. In some areas the hawthorn can be at risk of fire blight, so you need to be alert to this. The bushes are otherwise exceptionally robust and low maintenance.

Flavorsome cooking with hawthorn berries

Hawthorn purée with apples

Makes 2 × 1 lb (500 ml) jars

generous 1 lb (500 g) hawthorn berries
1 apple
generous ¾ cup (200 ml) apple juice
2¼ cups (500 g) gelling (jam) sugar
juice of 1 lemon
pinch of cinnamon

■ Wash the hawthorn berries and leave to drain. Peel and core the apple, then chop it into small pieces.

■ Place the hawthorn berries and chopped apple in a large pan. Add the apple juice, topping up with a little water so that the fruit is just covered.

■ Simmer for about 10 minutes until the berries are soft. Then pass through a sieve or food mill.

■ Return the fruit purée to the pan. Add the gelling sugar, lemon juice, and cinnamon and bring to a boil, stirring all the time. Allow to boil rapidly for about 4 minutes.

■ Pour into clean, sterilized jars and seal. Turn the jars upside down for a few minutes and then turn them the right way up to cool.

Fig chutney goes well with goat's milk cheese, other strong cheese, and grilled food.

Fig chutney with hawthorn berries

MAKES 3 × 8-OZ (250-ML) JARS

7 oz (200 g) hawthorn berries
1 vanilla bean
7 tbsp (100 ml) apple juice
14 oz (400 g) fresh figs
1 small apple
1 small shallot
1 piece of fresh ginger (about ½ in / 1 cm)
⅔ cup (150 g) brown sugar
pinch of cinnamon
pinch of ground cloves
½ tsp cumin
a few coriander seeds
½ tsp orange zest
salt and freshly ground black pepper
7 tbsp (100 ml) white wine vinegar

■ Pick any leaves or stalks from the hawthorn berries, wash them thoroughly, and leave to drain. Slit the vanilla bean in half lengthwise and scrape out the seeds. Place the vanilla seeds and bean in a bowl together with the hawthorn berries and apple juice, combine well, and leave to infuse overnight.

■ The next day, transfer the vanilla and fruit mixture to a large pan. Bring to a boil and simmer for about 10 minutes, until the berries are soft. Then pass through a sieve or food mill.

■ Wash the figs, rub dry, and cut them into chunks. Peel and core the apple and then cut it into pieces. Peel the shallot and the ginger and dice finely or grate them.

■ Bring the fruit purée to a boil together with the figs, apple, shallot, ginger, sugar, spices, and vinegar. Reduce the heat and leave to simmer for about 10–15 minutes, depending on the desired consistency.

■ Pour into sterilized, screw-top jars and seal. Turn the jars upside down for a few minutes and then the right way up again to cool. Store in a cool, dark place.

Rose hip—"little man in the red coat"

These berries can be seen brightening up hedges everywhere from September—"the little man in the bright red coat standing in the woods," to quote the German nursery rhyme. Regardless of whether they are dog roses, old garden roses, or specialist rose varieties, they all bear edible rose hips. The fruit can differ greatly in appearance, however: orange, pale, or dark red; a blue sheen imbued with black; or with a covering of tiny hairs. From oval or elongated, through onion-shape or bulbous, to round as a ball, these "little men" also come in all shapes and sizes.

Rose hips are among our most vitamin-rich fruits: their Vitamin C content is extraordinarily high, and they also contain vitamins B1, B2, B6, A, E, K, P (rutin) and the coloring agents, carotene and lycopin. Then there are the abundant minerals (including potassium, calcium, magnesium, and sodium) as well as malic and citric acids. This impressive range of essential ingredients makes the rose hip very successful at boosting the body's defenses. A mild infusion made from the dried, whole fruit has a stimulatory effect on the digestion as a whole and the kidneys in particular—so it should not be taken directly before going to bed. In the summer, this invigorating preparation is also great for drinking cold. The diuretic effect also has a positive impact on the joints and the ambulatory system. Although rose hips are high in carbohydrates, they are also high in water (more than 75 percent) and low in fat, which makes them ideal for healthy eating.

What is important is that the fruit is processed carefully in order to retain as many of the valuable nutrients as possible. Jams should be simmered gently and juices should not be heated too much (maximum 167 °F / 75 °C). To make a traditional fruit purée from rose hips, first remove the stalks and any remaining flower parts from the washed fruit. Then place the rose hips in just enough water to cover them and boil for about 30 minutes until they are very soft. Then you can purée them; pass them through a sieve or muslin cloth so that the seeds and any hairs are left behind. This purée can form the basis for a range of sweet or savory, passata-like sauces. Boiling the purée with gelling sugar, together with some red wine if desired, will produce the classic rose hip jelly. The seeds taste of vanilla and these, together with the leftover skins, can also be used; pour hot water over them and they will make a mild tisane. The fruit can be harvested until the first frost of the season—thereafter they become too mushy to process.

Rose hips are a valuable winter resource for birds; this is the reason for the introduction of the category "rose hip roses." This makes it easy for bird fans and berry gardeners to select the optimum variety for the garden. A number of varieties of the Gallic rose (*Rosa gallica*) are especially high-yielding—this applies to the Rugosa rose (*Rosa rugosa*) in particular. It bears very large fruit and is often planted along highways. It is the easiest variety to grow in the garden. The dog rose (*Rosa canina*) also produces a wonderful fruit harvest. The Burnet rose (*Rosa pimpinellifolia*) bears black fruit that produces a very dark purée. The "Pi lnitzer Vitaminrose" variety is, as its name suggests, particularly rich in vitamins.

Flavorsome cooking with rose hips

Rose hip jam

MAKES 3 × 11-OZ (300-ML) JARS

14 oz (400 g) rose hips
$^2/_3$ cup (150 ml) water
zest and juice of 1 unwaxed lemon
scant 1¼ cups (250 g) gelling (jam) sugar
4 whole cloves

■ Wash the rose hips and remove the stalks and remaining flower parts. Combine with the water and leave to infuse overnight.

■ The next day, transfer the rose hip infusion to a pan, bring to a boil, and then simmer over low heat for 30–45 minutes until the rose hips are soft. Purée and filter the mixture by passing it through a food mill (smallest hole setting) or muslin cloth.

■ Combine the rose hip purée with the lemon juice and zest, weigh, and place in a saucepan with the same weight in gelling sugar. Add the cloves, bring to a boil, and simmer for about 10 minutes, stirring all the time. Test to check if the setting point has been reached before removing the cloves.

■ Pour the boiling hot jam into hot, sterilized jars. Seal the jars, turn them upside down for a few minutes, and then turn them the right way up again to cool.

Rose hip and pear purée

MAKES 2 × 8-OZ (250-ML) JARS

generous 1 lb (500 g) rose hips
7 tbsp (100 ml) water
juice of 1 lemon
½ cinnamon stick
7 tbsp (100 g) runny honey
1½ lb (700 g) ripe pears
1⅓ cups (300 g) gelling (jam) sugar
⅔ cup (150 ml) dry white wine

■ Wash the rose hips and then remove the stalks and any remaining flower parts.

■ Bring the rose hips to a boil together with the water, lemon juice, and cinnamon, and simmer for 30–45 minutes until soft. Purée the mixture by passing through a food mill (smallest hole setting) or muslin cloth; this will also filter the seeds and hairs from the fruit. Stir in the honey and leave the purée to cool.

■ Peel and core the pears; then cut them first into quarters and then into cubes. Place in a pan together with the sugar and wine, cover, and simmer over medium heat for about 10 minutes until soft.

■ Purée the pears in a blender and stir into the rose hip purée. Place in bowls and serve. To keep the rose hip purée for longer, pour it into sterilized jars and seal.

The cornelian cherry—Turkish to Viennese delight

In 1683, the victorious troops of the Holy Roman Empire plundered the deserted camp of the besieging Ottoman army on the outskirts of Vienna. It was then that they discovered the famous sacks of coffee behind the burgeoning Viennese café culture. Less well known is the fact that they also found sacks full of cornelian cherry stones. Perplexed, they questioned their Ottoman prisoners as to the reason for these unusual spoils of war. The latter disclosed that adding the ground stones to the brewed coffee gave it a subtle vanilla flavor. This previously unknown refinement from the Ottoman tradition ultimately became the real secret behind the success of Vienna's coffeehouses.

But the cornelian cherry has an even longer history of diverse uses, including culinary, medicinal, and craftwork. Excavations have revealed that the fruit was eaten by our ancestors in the Early Stone Age. The bushes can reach 100 years of age, with a single specimen being sought out and harvested by successive generations. This wild fruit ultimately became so popular that harvesting was regulated by law. Its particularly hard, high quality wood was used to make a variety of tools and weapon handles.

Today, cornelian cherries (*Cornus mas*) are widespread throughout central and southern Europe. They are especially eye-catching in the early spring with their bright yellow, honey-scented clusters of flowers. These attractive, early blossoms are the reason why the

bushes are often planted in parks and gardens, but there are also wild specimens.

It was during my training in wild fruit cultivation trials that I became acquainted with the cornelian cherry. When fully ripe, the deep red fruit has an incomparable, full-bodied aroma. It is partly reminiscent of cherries—which, together with the red berry color, is how it got its name. It also has the subtle flavor of dark, ripe berries. I like this tender, melting fruit best in ice cream. Pure fruit jams made from cornelian cherries are also delicious, while a specialty from Austria is cornelian cherry Schnapps—the distilling process produces an outstanding concentration of aroma. A tasty savory pickle is made from the unripe (!) green-red fruit; these "wild olives" are harvested in August (they only ripen in September through October). They are then cooked in a little vinegar and water, together with spices such as peppercorns or garlic, for a few minutes. Placed in dark, airtight jars, they will keep for about a year.

For all other recipes fully ripe fruit are best, being the most flavorful. When ripe, the black-red berries quickly fall from the tree and need to be gathered from the ground as well. By this stage, you can be sure that the bitter tannins present in the unripe fruit have been converted into sugar. This eliminates the need for freezing, which can otherwise be used to achieve this transformation. Shaking the branches is not advisable as this could compromise next year's harvest. Intensive cross-breeding has increased the size of the fruit of varieties available today from what was originally a currant-size berry to an almost plum-size, pear-shape juggernaut (see Recommended varieties on p. 179). The plant itself is a low maintenance, large, sprawling bush that grows to a maximum height of 26 feet (8 m). In the wild, its preferred locations are forest glades and hedges in warmer, drier areas. The plant does benefit from regular watering when the fruit are ripening, however—otherwise the fruit can drop prematurely.

Flavorsome cooking with cornelian cherries

Cornelian cherry cake

MAKES 1 SPRINGFORM PAN (10 IN / 26 CM DIAMETER)

generous 1 lb (500 g) cornelian cherries
butter and flour for the pan
generous 1 cup (250 g) soft butter
scant 1 cup (200 g) sugar
pinch of salt
3 tsp baking powder
2 tbsp cocoa powder
5 tbsp (40 g) poppy seeds
2 2/3 cups (400 g) all-purpose flour
5–6 eggs
4 tsp (20 ml) cherry Schnapps
5 tbsp milk
confectioners' sugar for dusting

■ Pick any leaves or stalks from the cornelian cherries, rinse in hot water, and pat dry. Grease a springform pan with butter and dust with flour.

■ Pre-heat the oven to 350 °F (180 °C) (fan). Beat the butter with the sugar and salt until creamy.

■ Stir in the baking powder, cocoa powder, poppy seeds, and flour. Gradually stir in the eggs. Add the cherry Schnapps and milk and mix to a smooth dough.

■ Place the mixture in the springform pan, spread the cornelian cherries on top, and press them in slightly. Bake in the pre-heated oven for 45–50 minutes.

■ Allow the cake to cool slightly before removing from the springform pan. Place on a wire rack to cool.

■ Dust with confectioners' sugar and serve. Provide a small bowl for the stones.

Cornelian cherry ice cream

Serves 4

generous 1 lb (500 g) cornelian cherries
$^2/_3$ cup (150 ml) cold water
4–5 tbsp confectioners' sugar
6 tbsp (100 g) cream
mint leaves for the garnish

■ Pick any leaves or stalks from the cornelian cherries and wash them in hot water. Transfer the cherries to a pan, add cold water as per the ingredients list, and boil until the fruit is soft. Remove the stones by hand or by passing through a sieve or food mill.

■ Combine with the confectioners' sugar and the cream, adding more sugar if required. Place in the freezer for about 3 hours, or finish in an ice cream maker.

■ Remove from the freezer about 10 minutes prior to serving and leave to thaw slightly. Serve in individual bowls, garnished with fresh mint leaves if desired.

Cornelian cherry jam

MAKES 2 × 8-OZ (250-ML) JARS

14 oz (400 g) cornelian cherries
²/₃ cup (150 ml) apple juice
generous 1 cup (250 g) gelling (jam) sugar
1 vanilla bean

■ Pick any leaves or stalks from the cornelian cherries, wash in hot water, and simmer in a pan with the apple juice until the fruit is soft.

■ Then pass the cherries through a sieve or food mill. Weigh the fruit purée and combine with an equal quantity of gelling sugar.

■ Slit the vanilla bean in half lengthwise, scrape out the seeds, and stir into the cornelian cherry purée. Bring to a boil, stirring all the time, and simmer for about 6 minutes before testing for the setting point.

■ Place in clean, sterilized jars and seal. Turn the jars upside down for a few minutes and then the right way up again to cool.

Sloeberries—the wild, thorny ones

Wild and thorny are what they are, these 6–16 feet (2–5 m) high, branching trees with their black wood. The blackthorn (*Prunus spinosa*), commonly known as sloe, is widespread as a hedging plant. Anyone wanting to harvest its purplish-black fruit has to fight through the thorns as in a fairytale to get to it. And what can we expect then? No sweet, juicy reward for all the effort—just a tart, little, wild plum known as the sloeberry. So what exactly makes the sloe so indispensable that it has been gathered in large quantities since the Stone Age?

Sloes contain large quantities of tannins, which are what cause the lips to pucker. Chemically speaking, they have an astringent effect, drying out the mouth and throat. Timing of the harvest is key—the bitter tannins break down almost completely after the first frosts in October / November. Only then does the sloeberry's sugar content come to the fore (up to 10 percent). Now you will be able to detect the fine sloe aroma and taste the difference over its big sister, the plum.

If you cannot wait until November—or if you want to get in ahead of the birds—you can also freeze the fruit, several times consecutively. I recommend cutting off whole branches to avoid having to pick every single sloeberry individually from the thorns. Freeze them as they are and then simply shake the berries off. This method damages neither the fruit nor your fingers. It is

suitable only for your own, homegrown plants, however, because according to nature conservationists and environmentalists, the wild sloe provides a habitat for numerous animals and should not be harvested.

The high Vitamin C content in particular (up to 50 mg per 3½ ounces / 100 g of fruit) together with anthocyanins, B vitamins, minerals, and tannins make the sloeberry a winter-time vitamin bomb. I like the fruit best preserved, in sweet-and-sour pickle, or with winter spices in red wine. The robust flavor goes wonderfully well with game dishes. The aromatic acids harmonize very well in mixed fruit jams with apples, pears, and quinces. Traditional sloe recipes include juice, wine, and liqueur. The white blossoms (March/April) can also be used to make a liqueur with a characteristic, bitter almond aroma. This is especially recommended as a dessert sauce for ice cream and other sweet dishes. Similar to the popular slivovitz, which is distilled from plums, there is also a Schnapps distilled solely from sloeberries. This is a particularly noble way of preserving the fine aroma.

The gentle steam juicer process extracts some of the bitter almond aroma from the stone of the fruit and imbues the juice with it. When processing sloeberries you should take care to remove the stones as they do contain small quantities of prussic acid—as do many other stone fruit varieties. Small amounts, such as those devolving from the production of alcohol, for example, are harmless. Pure sloe juice is used for detoxifying and vitality regimes in the spring in particular. Traditionally, sloes have been used to boost the metabolism, for overall strength (e.g. during convalescence), and to strengthen the heart.

Sloes have only a limited suitability for the garden. It is a tree that grows slowly, but broadly. With its shallow roots, the sloe can access even steep slopes and rocky outcrops with its extensive offshoots. This makes it suitable only for larger gardens and tolerant neighbors. It is winter hardy up to –16 °F (–27 °C) and so the sloe flourishes even in places where practically no other fruit growing is possible. There are large-fruited varieties devolved from cross-breeding with true plums, which do not spread quite as vigorously. A particularly attractive variety is the "Purpurea"—its pink blossoms hover like a cloud over the red leaves.

Flavorsome cooking with sloeberries

Sloe gin

MAKES 1½ QUARTS (1.5 LITERS)

generous 1 lb (500 g) sloeberries
²/₃ cup (150 g) sugar
3 whole cloves
1 cinnamon stick
1 quart (1 liter) of gin

■ Pick any leaves or stalks from the sloeberries, wash them, and pat dry. Place them in a large (1½ quarts / 1.5 liters) preserving jar with the sugar, cloves, and cinnamon stick.

■ Fill with enough gin to cover all of the ingredients and seal well. Leave to infuse in a dark, warm place for 8–10 weeks, turning the jar occasionally.

■ Then strain through a muslin cloth or fine sieve and store the sloe gin in clean, sterilized bottles.

Sloe jam

MAKES 2 × 8-OZ (250-ML) JARS

1¼ lb (600 g) sloeberries
1 unwaxed lemon
generous 1 cup (250 g) gelling (jam) sugar

■ Pick any stalks or leaves from the sloeberries, wash them, and then cook them in a little water until soft, stirring occasionally and adding a little more water if necessary.

■ Then pass through a sieve or food mill. This will produce about 1 lb (500 g) of sloe purée.

■ Grate the zest of the lemon and then squeeze out the juice. Place in a pan with the sloe purée and gelling sugar and bring to a boil, stirring all the time. Boil rapidly for about 4 minutes, stirring frequently.

■ Pour the jam into hot, sterilized jars and seal. Turn the jars upside down for a few minutes and then turn them the right way up to cool.

Snowball berries—from the country garden

You will find them in many ornamental and country gardens—those huge, white balls of blossom like snowballs (*Viburnum opulus*). They stand there so decoratively, charming the eye of the passer-by. There are no berries to be seen, however; only red leaves adorn this attractive bush in the fall. On the edge of the forest, though, the original, wild variety grows; it bears large, white, plate-size blossoms from May through August. Then the berries appear in heavy clusters, and are ready for picking from October. Its similarity to the elder accounts for the other name, water elder, for this tall plant with oak-like leaves that can grow up to 20 feet (6 m) in height. As with the elder, the unripe fruit are slightly toxic; they do need to be properly ripe for eating! You should also refrain from eating the berries raw. Harvested after the first frost and cooked for 5–10 minutes, however, the last of the toxins disappear from the fruit and the result is a highly aromatic fruit purée. The tart taste is highly reminiscent of lingonberries, with which snowball berries combine very well.

Snowball berries are for fans of robust, rustic flavors. They are a new discovery even in sophisticated kitchens. Anyone wanting to snack on just a few berries from the bush will be stopped in their tracks before they can do so, however. The intense valerian and—my apologies, but I have to say it—sweaty feet aroma is a deterrent, certainly at first. But with the effects of frost and heating, the smelly and slightly toxic components (viburnum, valerenic acid, and butyric acid) dissipate.

To harvest the fruit, cut the berry clusters off cleanly, without stems if possible. Cook the washed fruit with a little water and then pass through a fine sieve or food mill. This removes the heart-shape stones and produces a fruit purée, which is outstanding as a jam in combination with other fall berries. I especially like it in pickle as a wonderful accompaniment to a strong cheese or venison roast. It is also a treat on bread, with good butter, a slice of cold, roast meat, and snowball chutney.

In Turkey, the berries are used to make a refreshing drink called "Gilaboru." In Latvian folk medicine they are used to heal burns, while in Norway and Lithuania an infusion of fresh snowball berries, blossoms, and honey is used to treat colds and coughs. Russian studies have shown the berries to be beneficial for diabetics. They contain little sugar, but have plenty of pectin, tannins, essential oils, amino acids, and vitamins A, C, and K, as well as flavonoids, acetic acid, and valerenic acid. It should be noted, however, that their antispasmodic properties make them unsuitable for consumption during pregnancy.

In the garden, the snowball is widely popular as a bush or hedge. It prefers the semi-shade and a location that is not too dry. As it attracts aphids, resourceful gardeners plant it at the edges of their gardens. The large-fruited varieties are particularly easy to propagate from cuttings.

Serviceberries—a delight for the eyes and palate

The wild serviceberry (genus *Amelanchier*) was discovered by winemakers early on—as a fruity snack in sunny vineyards. Edible when raw, the very sweet fruit were a quick source of renewed energy in the form of fructose during the hard physical labor of grape picking. In the mountains of central, southern, and eastern Europe, this multi-stemmed bush grows to up to 8 feet (2.5 m) in height and is often to be found on heat-storing, rocky outcrops. It can also be found in forest glades, sunny spots at the edge of forested areas, and parkland.

From April through May, the abundant blossoms cover whole slopes in white. The deep violet, apple-shape, pea-size berries are ripe for picking from July onward. The bush again becomes a feast for the eyes in the fall: with its orange through scarlet leaves, in the USA where it is also known as the Juneberry or shadblow, the serviceberry is an integral feature of Indian summers. It therefore comes highly recommended for the home yard or garden, as well as for balconies and terraces. Its popularity is evident in its many common names in other languages—Edelweiß tree or raisin tree, chamois berry, and bell berry. For me, the serviceberry is one of the rediscoveries of the last few years.

The sweet fruit is in demand not only in Europe; in Native American cuisine, the berries of sister varieties (*A. alnifolia*, *A. canadensis*, *A. laevis*, *A. lamarckii*, *A. spicata*) are used in a manner similar to

raisins, particularly for the winter-time energy dish pemmican. This mixture of berries, dried buffalo or venison meat, and fat keeps for the whole winter and can be eaten both raw and cooked.

As a passionate "grazer gardener" I enjoy the bilberry-like flavor of the serviceberry with its slight marzipan nuance. But if the freshly picked fruit does make it into my kitchen, I like to turn it into a refreshing summer smoothie. Mixed with apple or pear juice, serviceberries make an especially delicious drink. They are also a great addition to red berry jellies or in mixed fruit jams. As they are relatively high in pectin, you do not need to use as much gelling sugar when processing them.

In the case of illicit or even unbridled berry snacking, I do recommend washing your hands thoroughly afterward as, like wild bilberries, serviceberries do leave blue traces on pilfering fingers. These coloring agents belong to the anthocyanin group and as free radical neutralizers are highly beneficial to our bodies: they are antioxidants and boost the immune system. Serviceberries are rich in vitamins and contain plenty of skin nurturing potassium. The trace elements iron, calcium, magnesium, and phosphate make serviceberries thoroughly healthy. Their tannins help to heal throat infections and can be utilized to make flavorful juices and liqueurs.

Like peach and cherry pits, the tiny seeds inside the fruit contain low quantities of prussic acid. This only becomes a concern if you eat very large quantities of chewed seeds, though. Only in such

cases are stomach and digestive complaints likely. Whole seeds remain undigested and harmless and are simply eliminated from the body.

I have now planted many different varieties of serviceberry in my garden. This robust, low maintenance, wild berry bush is a worthy addition to any edible hedging plants. Its natural habitat makes it suitable for particularly dry spots in the garden, such as a newly planted slope, or in front of a hothouse. The berries will be especially sweet and tasty here. Regular feeding with organic fertilizer ensures a strong fruit set. The berries do not all ripen at the same time, but rather in succession. This means that you have something available to snack on over a longer period of time. Do remember, though—birds like serviceberries too!

The North American varieties are significantly larger, growing up to 26 feet (8 m) in height, and so they do need plenty of space between them and the other plants in the hedge. With the umbrella-like spread of their branches, they form a summery blossom canopy ideal for creating a romantic seating area. These varieties, too, are very undemanding, but they do not like it quite as dry. None of the serviceberry varieties is particularly fond of heavy soil, which will need to be thoroughly loosened. The smaller varieties are ideal for balcony gardeners and, with adequate watering, will bear lots of fruit even when grown in a container.

My recommended varieties: p. 181.

Flavorsome cooking with serviceberries

Serviceberry and rose jam

MAKES 4 × 8-OZ (250-ML) JARS

2¼ lb (1 kg) serviceberries
2¼ cups (500 g) gelling (jam) sugar
1–2 handfuls fresh rose petals
2 tbsp lemon juice

■ Wash the serviceberries, remove them from their stalks, and leave to drain well.

■ Combine the gelling sugar with the rose petals and use either a food processor or handheld blender to purée the mixture. Mix with the serviceberries and lemon juice, cover, and leave to infuse overnight.

■ The next day, transfer the infusion to a large pan and quickly bring to a boil. Continue to boil for about 4 minutes, stirring all the time. Test for the setting point and boil for a little longer if required.

■ Pour into hot, sterilized jars and seal. Turn the jars upside down for a few minutes and then the right way up to cool.

Serviceberry clafoutis

Serves 4

14 oz (400 g) serviceberries
4 eggs
4 tbsp confectioners' sugar
6 tbsp all-purpose flour
scant 1 cup (225 ml) milk
butter for greasing
sugar for dusting

■ Pre-heat the oven to 430 °F (220 °C) (fan). Wash the serviceberries and pat them dry.

■ Beat the eggs with the confectioners' sugar until foamy, then fold in the flour and milk.

■ Grease four round, ovenproof dishes (around 2–3 inches/ 6–8 cm in diameter) and divide the serviceberries equally between them. Pour the egg custard over them, smooth the surface, and bake in the pre-heated oven for about 25 minutes until golden brown.

■ Leave to cool slightly before serving, still lukewarm in their dishes and dusted with sugar.

Silverberries—decorative and aromatic

These attractive plants from Central Asia have been grown as ornamental bushes or small trees in parks and gardens since the 17th century. Their decorative foliage has a silvery-green sheen. The blossoms give off a wonderful, honeylike fragrance, which appeals strongly to both man and bee. Silverberries (genus *Elaeagnus*) are even to be found on the grass verges bordering highways, thus demonstrating their robustness, modest requirements, and winter hardiness. What very few people know, however, is that the berries are good for you and are highly aromatic, fruity, and bittersweet to the taste. They are easy to harvest and can be used to create all kinds of culinary delights.

The varieties of culinary interest include the oleaster (*Elaeagnus angustifolia*) whose green-gray foliage with its distinctive, silvery shimmer makes this arching bush a visual delight. The inconspicuous but wonderfully scented blossoms turn into yellowish through orange fruit from July. They have a full, fruity flavor and, when dried, a nutty aroma. They are a staple food in Asia. The fruit from cross-bred varieties can be up to 1 inch (2 cm) in length and ½ inch (1 cm) thick. In addition to their high protein content (10–55 percent), they are also rich in glucose, fructose, potassium, and phosphorus. They are processed like the fruit of the more widespread cherry silverberry (*Elaeagnus multiflora*).

The white blossoms have a distinct honey fragrance and are used for scented potpourris and perfumes, while the orange-red round or elongated fruit is a particular delight. When fully ripe, the berries are very juicy and have a refreshing, bittersweet aroma. They are very popular in China, Japan, and Korea as a fruit snack. They are cooked and processed into a purée or juice with a pleasant sweetness and wonderful, rich red color. They are also to be recommended for berry jams or jellies. They are delicious in combination with quark, yogurt, and cream, as ice cream or as a milkshake. In sweet dishes, they combine well with the related sea buckthorn. Their aroma is also ideal for alcoholic beverages. I am especially fond of them as a refreshing sorbet.

The third variety that is of culinary interest is the Japanese silverberry (*Elaeagnus umbellata*), which comes from North East Asia. It is very similar to the cherry silverberry in appearance and also has silvery green foliage. It is exceptionally frost hardy—the "Red Cherry" and "Sweet Scarlet" varieties with large red berries bred in Ukraine have now come onto the market. There, they are extensively grown on plantations and regularly produce high yields. Once they have reached optimum ripeness, which happens between September and November, the fruit can simply be shaken from the branches. This smaller variety does not form offshoots, which makes it suitable for small yards and patio containers. The berries are processed as for the cherry silverberry (see above).

All silverberry varieties are easy to care for and low maintenance. They can also be planted on embankments, roof overhangs, or gravel surfaces where, reaching up to 13 feet (4 m) in height, they are significantly more robust than the frequently planted cherry laurel. They are also highly suitable for inner city areas or balcony gardens. They provide a good harvest even in difficult, barren locations—a fruit for lazy bed gardening.

Medlars—headstrong, with character

Could the Romans have anticipated the long culinary tradition they were founding when they brought the medlar (*Mespilus germanica*) to central Europe on their conquests? From times of Antiquity through the Middle Ages, no monastery or country garden was complete without this highly sought-after plant. Growing as a bush or small tree, the medlar with its downy, yellowish-brown fruit was a common sight for centuries. St. Hildegard von Bingen recommended its fruit for blood purification and an overall boost, particularly for the frail. Whole medlar gardens were planted, which from 1492 were even protected by law. At this time, the bushes still had huge thorns on the short, fruit-bearing branches, and their small berries were usually eaten raw. Cultivation went into decline from the 17th century onward, however; the variety regressed and ran wild. Medlars can still be found growing wild today, but better suited to home growing are the cultivars with no thorns and larger fruit. The medlar bush is a sensory delight with its characteristic, often picturesque shape, large, white blossoms, yellowish-orange fall foliage—and aromatic fruit.

So what is it that makes this hard and relatively unattractive fruit so special? Biting into it in the fall will pucker your mouth instantly. It is only after a few night frosts that the berries part with their secret; that is when the flesh turns brown, aromatic, and meltingly soft. There is no trace of the tannins and fruit acid, these now having transformed themselves into sugar. If the frost is late in coming, you can also freeze the fruit or store it for at least four weeks in moist bran or sawdust. The medlars will then be so soft that you can eat them directly with a spoon. The slightly acidic, winy flavor and apple-almond aroma of the tender fruit is a fascinating combination; it is not to everyone's taste, but does have its passionate followers.

It is in jams, either on its own or mixed with other wild fruits, that the medlar displays the best of its culinary attributes. The high pectin content makes processing easier and reduces the need for added gelling agents. The cooked fruit flesh is best passed through a muslin cloth or food mill in order to remove the seeds and skin. With the large fruit varieties available today, the seeds are generally easy to remove. I like to prepare the fruit in the same way as baked apples: I remove the cores and fill them with a combination of almonds, raisins, and brown sugar. Then I place them in the oven. Drizzled with medlar Schnapps and served with a scoop of vanilla ice cream they are a true delight!

When planting them in the garden you do need to give each medlar at least 13 feet (4 m) of space, as the broad growth of each bush means they will become entangled. The longer branches need to be pruned back occasionally, as the medlar blossoms and bears fruit on its shorter branches. A warm, sunny spot will ensure that the fruit ripens well. A wonderfully special variety is the "Kurpfalz medlar," although it is only propagated by a few nurseries. Its fruit contains no tannins and can be enjoyed directly from the tree like apples—no need for frost or storage time.

Flavorsome cooking with medlars

Medlars in dressing gowns

Serves 4

1¾ cups (250 g) all-purpose flour
8½ tbsp (125 g) butter
2 eggs
9 tbsp (120 g) sugar
confectioners' sugar for dusting
8–12 medlars
mint leaves for the garnish

■ For the dough, quickly work together the flour, butter, 1 egg, and the sugar. Shape into a ball, cover in plastic wrap, and chill in the refrigerator for 1 hour.

■ Pre-heat the oven to 350 °F (180 °C) (fan). Roll out the dough as thinly as possible without it tearing. To do this, it is best to roll it out between two layers of plastic wrap.

■ Cut into 8–12 large, equal-size squares, according to the number and size of the individual fruit, with one medlar fitting each square. Dust the squares lightly with confectioners' sugar. Place a washed and dried medlar in the middle of each square, fold the dough into a parcel around it, and squeeze the edges together.

■ Whisk the other egg and use to brush the medlar parcels. Place the medlars in their "dressing gowns" on a baking sheet lined with wax paper and bake in the oven for about 30 minutes, until golden-brown.

■ Place the medlars on plates, dust with confectioners' sugar, and garnish with mint leaves. Serve with a scoop of vanilla ice cream if desired. Provide a small bowl for the medlar seeds.

Medlar and peach compote

Serves 4

6 ripe peaches
generous 1 lb (500 g) ripe medlars
¾ cup (200 ml) red wine
3½ tbsp (50 g) sugar
1 tbsp vanilla sugar
¼ handful mint leaves

■ Wash the peaches, cut them into halves or quarters, and discard the stones. Wash the medlars.

■ Bring the wine to a boil together with the sugar and vanilla sugar. Add the fruit and simmer gently for about 5 minutes until you have the desired consistency.

■ Remove the mint leaves from their stalks, wash, pat dry, and fold into the compote.

■ Place the compote in a large serving bowl and serve as an accompaniment to ice cream or grilled dishes. Provide a small bowl for the medlar seeds.

Harvest calendar

Barberries	October–November
Bilberries	July–September
Blackberries	July
Chokeberries	August–September
Cornelian cherries	September–October
Cranberries	September–November
Currants (black)	July
Currants (red, white, green)	June–August
Elderberries	September–October
Figs	September–October
Flowering quince	October–November
Fourberries	June–July
Goji berries	August–September
Gooseberries	June–July
Hawthorn	September–October
Japanese wineberries	July–September
Josta berries	June–July
Juniper	all year
Kiwis	October–November
Lingonberries	September–October
Mahonia	October–November
May berries	May
Medlars	November–December
Raspberries	June–September
Rose hips	September–November
Schisandra	September–October
Sea buckthorn	September–October
Serviceberries	July–August
Silverberries	September–November
Sloeberries	October–November
Snowball berries	October–November
Strawberries	May–June
Table grapes	September–November

Recommended varieties

Barberries
Alba (white fruit), Dulcis (red, less acidic).

Bilberries
American blueberries: Bluecrop, Bluetta, Bluegold, Bluejay, Brigitta, Chandler, Darrow, Duke, Earlyblue, Elliot, Goldtraube, Herbert, Lateblue, Nelson, Northland, Patriot, Reka, Rubel, Sierra, Sunrise, Toro.
Dwarf varieties: Ama, Berkeley, Coville, Dixi, Heerma, Jersey, Spartan, Top Hat.
Red-fruited: Red Winner.

Blackberries
Asterina, Black Satin, Chester/Thornless, Cutleaf Blackberry, Hull Thornless, Jumbo, Loch Ness, Loch Tay, Mammoth, Navaho, Tayberry, Theodor Reimers, Thornfree, Thornless Evergreen, Triple Crown, Wilsons Early.
Fall blackberries: Primocane Reuben.

Chokeberries
Smaller varieties for the home garden (up to 5 feet / 1.5 m): Aron, Hugin, Nero and Viking.
Up to 6½ feet / 2 m high and strong colorant: Rubina.
Simultaneous ripening, high yield: Galicjanka.

Cornelian cherries
Large fruit or high-yielding: Cormas, Devin, Hecoma, Jolico, Kasanlaker, Schumener, Schönbrunner Gourmet Dirndl, Titus.
As male plant for guaranteed pollinating and pollen supply: Mascula.
For home gardens and wild fruit hedges, resistant to pruning and quick to regenerate: Aurea (yellow leaves), Variegata (colored foliage), Alba (white fruit), Flava (yellow fruit with a "wild" flavor).
New Hungarian and Polish varieties (red and yellow): Bolestraszycki, Cornello, Dublany, Ekzoticznyj, Elegantnyj, Jantarnyj, Juliusz, Koralłowyj Marka, Paczoski, Swietljaczok, Szafer, Wydubieckij.

Cranberries
For the hobby gardener: Bain McFarlin, Beckwith, Bergmann, Black Veil and Searles.
The most important commercial varieties worldwide include: Ben Lear, Big Four, Early Black, Howes, Jumbo, Pilgrim, Red Balloon, Top Hat.

Currants, black
Andega, Ben Alder, Ben Connan, Ben Lomond, Ben Moore, Ben Sarek, Ben Tirran, Ben Tron, Byelorussian Sweet, Bona, Cassimia Neva, Cassimia Nimue, Ceres, ECM, Farleigh, Fertöder 1, Foxendown, Goliath, Hedda, Intercontinental, Kieroyal, Kristin, Leandra, Langtraubige Schwarze, Malling Jet, Narve Viking, Noir de Bourgogne, Öjebyn, Ometa, Polar, Silvergieters, Roodknop, Rosendahls, Silvergieters Schwarze, Stroklas, Tenah, Tifon, Titania, Tsema.
New Cassissima series: Neva, Nimue, Noiroma.

Currants, red
Detvan, Erstling aus Vierlanden, Heinemanns, Heros, Houghton Castle, Jola, Jonkheere van Tets, Junifer, Red lake/Roter See, Ribest Babette, Ribest Lisette, Ribest Violette, Rolan, Rondom, Rodneus, Rosa Holländer, Rote Holländische, Rote Versailler, Rosalin, Rosetta, Rotet, Rote Vierländerin, Rovada, Stanza, Tatran, Telake, Traubenwunder.

Currants, white
Blanka, Primus, Ribest Blanchette, Vitjätte, Weiße Holländer, Weiße Jüteborger, Weiße Langtraubige, Weiße Versailler.

Elder

Especially large fruit: Haschberg, Mammut, Riese aus Vossloch, Sampo.
Very good flavor e.g. for jams: Fructo Lutea (pale fruit), *Sambucus nigra* var. *albida* (white), and *Sambucus nigra* var. *viridis* (green).
Red-leaved: Black Beauty, Black Lace, Black Tower, Guincho Purple, Thundercloud.
For dark, shady corners of the garden, with colored and/or golden foliage: Aurea, Madonna, Pulverulenta.
For container gardeners: Pygmea (8 inches / 20 cm high), Pyramidalis (conical in shape).

Elderberry, red

Anna, Plumosum Aurea, Sutherland Gold.

Figs

Bayernfeige/Violetta, Brown Turkey, Brunswick/Braunschweig, Contessina, Dalmatie, Dauphine, Fehmarn, Goldfeige, Goutte d'or (Doree), Hardy Chicago/Mongibello, Longue de Aout/Bananenfeige/Jerusalem, Pastiliere, Pfälzer Fruchtfeige, Ronde de Bordeaux, Nordland Bergfeige, Schweizer Bergfeige.

Flowering quince

Apple Blossom, Boule de Feu, Cido, Crimson and Gold, Friesdorfer Orange, Gold Calif, Knap Hill Scarlet, Moerlosii, Nicoline, Nivalis, Pink Lady, Rowallane, Rubra Grandiflora.

Fourberry

Black Gem, Black Pearl, Black Saphir.

Goji berries

Big Lifeberry, Lubera Instant Success, Natascha, Sweet Lifeberry, Synthia.

Gooseberries

Achilles, Captivator, Crispa Flavia, Crispa grün, Crispa Solemio, Espera, Früheste Gelbe, Grüne Flaschenbeere, Grüne Riesenbeere, Hinnomaeki yellow, Hinnomaeki green, Hinnomaeki red, Invicta, Karlin, Larell, Maiherzog, Mucurines, Pax, Redeva, Reflamba, Relina, Remarka, Resistenta, Roko, Rokula, Rote Eva, Rote Orleans, Rote Triumph/Triumphbeere, Rixana, Spinefree, Tatjana, Thumper, Weiße Triumphbeere, Xenia.

Josta berries

Jocheline, Jochina, Josta, Jostine, Jogranda, Rikö.

Kiwis

Actinidia deliciosa: Abbot, Bruno (small, cyclindrical, very aromatic fruit, ripe for eating shortly after harvesting, 400–600 fruit per bush), Green Light (only female, ripens very early), Hayward (female, large fruit, good appearance and flavor, ripe for eating from January to April), Kiwigold (female, yellow flesh), Jenny (self-pollinating, tried and tested, walnut-sized), Matua (male pollen source), Monty, Starella (only female).
Actinidia arguta: Ambrosia (tried and tested, large fruit), Ambrosia Grande (improvement on Ambrosia, very large fruit, early), Ananasnaja (very robust), Bayernkiwi, Bayerwald, Bojnice (apple-shaped), Geneva (group of Swiss varieties), Issai (tried and tested, self-pollinating), Julia ("Sachsenkiwi"), Kens Red, Kiwino (Japanese variety), Nostino (male pollinating variety), Maki/Amadue (only female, red fruit), Purpurna Sadowa (red fruit, Ukrainian), Rote Potsdamer (over one hundred years old, from the Sanssouci Park in Potsdam), Weiki (very frost hardy, vigorous growth).

Lingonberries

Red Pearl, Erntedank, Erntesegen, Koralle.

Mahonia

Robust, up to 5 feet/1.5 m high: Mirena, Jupiter, and Pamina (high-yielding hybrid, especially mildew-resistant).
Shorter varieties with deep red leaves: Apollo, Atropurpurea, Brilliant.

May berries

From Switzerland: Maistar, Mailon.
From the USA: Berry Blue, Blue Bird, Blue Belle, Blue Velvet.
Also Azur, Czelabinka, and the new ones Balalaika, Polar Bear, and Kalinka. Amur, Balalaika, Fialka, Maja, Morena.

Medlar

Oldest varieties still commercially available: e.g. Dutch Large-Fruited, Metz Medlar, and Nottingham Medlar. Sweet medlar (Kurpfalz medlar).
Especially large-fruited: Apyrena, Bredase Reus, Delice des Vannes, Dutch, Evreinoffs Monströse, Early English, Kings Medlar, Krim, Macrocarpa, Royal, Seedless, Hungarian, Westerveld.

Raspberries

Summer varieties: Back Jewel (black), Dr. Bauers Rusilva, Cumberland (black), Yellow Antwerper, Glen Ample, Golden Everest (yellow), Golden Queen (yellow), Himbo Star, Malling Exploit, Malling Promise, Marla, Meeker, Niniane, Preussen, Rubaca, Sanibelle, Schönemann, Tulamagic, Tulameen, Schöneman, Wei-Rula, Williamette, Winklers Sämling, Zefa 1 and Zefa 2.
Fall varieties: Autumn Amber (yellow), Autumn Belle, Autumn Best, Autumn Bliss, Autumn First, Autumn Sun, Aroma Queen, Fall Gold (yellow), Fall Red, Himbo Top, Korbfüller, Lloyd George, Pokusa, Polka, Romy, Saxa Record, Zefa 3, Twotimer.
Repeat flowering: Sugana yellow, Sugana red.

Rose hips/Roses

PiRo 3 Vitaminrose, *Rosa gallica*, *Rosa rugosa* (rugosa rose), *Rosa villosa*, and "rose hip rose" varieties.

Schisandra

Take 5.

Sea buckthorn

Female varieties with high active agent content: Askola, Dorana, Friesdorfer Orange, Frugana, Hergo, Julia, Leikora, Pendulina, Orange Energy, Romeo and Sandora.
Good male pollinating varieties: Hikul and Pollmix 1–4 (blossom over a longer period in order to pollinate all female varieties).

Serviceberries

Pacific serviceberry or saskatoon (*Amelanchier alnifolia*). Varieties: Ballerina (larger fruit), Rubescens (pinkish blossoms), Prince William, Bluemoon, Diana, Edelweiß, Forestburg, Obelisk (slim), Saskatoon Saskablue, Smoky, Dwarf Serviceberry (*Amelanchier ovalis* ssp. *pumila*).

Sloeberries

Large-fruited varieties: Godenhaus, Nittel, Plena (full blossoms, often produces double fruit), Purpurea (pink blossoms, few thorns, red leaves).

Snowball berries

Yellow leaves: Aureum, Harvest Gold, Park Harvest, Cranberrybush, Anny's Magic Gold.
Colored leaves: Variegata.
Dwarf form: Nanum.
Yellow-fruited hedge plants: Xanthocarpum, Roseum (sterile).

Strawberries

Single crop: Amazone, Asia, Avalon Classic, Avalon Delizia, Daroyal, Deutsch Evern, Elsanta, Erdbeerwiese, Florence, Fraroma, Gariguette, Georg Soltwedel, Gorella, Great Blossom, Hansa, Honeoye, Jucunda, Kaisers Sämling, Korona, Madam Moutot, Malwina, Mieze Nova, Mieze Schindler, Nerina, Oberschlesien, Osterfee, Polka, Regina, Simica, Sonata, St. Pierre, Thuchief, Thulana, Thuringia, Senga Sengana, Viba santa, Wädenswil 6, Weiße Ananas, White Dream, Wunder von Köthen.
Repeat crop: Elan, Fruirose, Mara de Bois, Josee, Ostara.
Wild strawberries: Alexandria, Baron von Solemacher red/white, Red Wonder, Rimona Hummi, Rügen, Siskeep, Yellow Wonder.

Table grapes

Only fungus-resistant PiWi varieties or Robusta varieties: Arkadia, Aromata, Birstaler Muskat, Boscoso, Exelsior, Fiorito, Frumoasa, Glenora, Glory Robusta, Himrod, Kodrinaka, Lakemont, Lilla, Muscat bleu, Nero, Ontario, Phoenix, Porza, Romulus, Rosolio, Russata, Solara, Suffolk, Tonia, Vanessa, Venus.

Index

Page numbers in italics refer to use in recipes.

Berries

Recipes

Useful addresses:

UK: www.thompson-morgan.com; www.marshalls-seeds.co.uk; www.primrose.co.uk; www.scotplantsdirect.co.uk/; https://www.rhs.org.uk/Advice/Grow-Your-Own/Fruit

US: www.penseberryfarm.com; www.buygojiberryplants.com; www.starkbros.com; www.noursefarms.com; www.backyardberryplants.com; http://www.agmrc.org/commodities__products/fruits/

Image credits

Leo Baranchuk-Chervonny: p. 106

Flora Press: pp. 155, 167, 172 right, 174 right

Fotolia: pp. 20 left, 21, 24, 26 both, 27, 30 small left, small middle, 32 both, 33, 35, 36 left, 39, 44 both, 46, 47, 48 left, right, 54 both, 55, 56 left, 58, 59 both, 62, 68, 70 left, 74 both, 78 left, 82 left, 84, 85 both, 86 left, 87 right, 89, 91 right, 94, 95 right, 98 both, 99, 101, 102 left, 107, 110 right, 114, 132 left, 133 right, 136 both, 138 right, 140 left, 142 both, 146 both, 150 both, 154 both, 160 both, 164 right, 165, 166 both, 168, 169, 170 left, 174 left

Christian Havenith: p. 8 top

iStockphoto: pp. 5 left, 13, 17, 19, 23 top right, 70 right, 71, 80, 88, 102 right, 104 right, 110 left, 124, 164 left, 178 left, 178 right, 179

Bildagentur Look: pp. 38, 40 right, 120 small middle, 181

Christine Paxmann: pp. 16, 18

Roger Prat: p. 172 left

Shotshop: p. 29

Stockfood: cover 1, cover 4 all, pp. 4 both, 5 right, 6, 8 below, 9, 10 all, 12, 14, 20 right, 23 top left, below left, below right, 25, 30 large, small right, 36 right, 37, 40 left, 41, 42 both, 43, 45, 48 middle, 51 all, 53, 56 right, 57 both, 60, 61, 63, 65, 66, 67, 69, 72 both, 75, 77, 78 right, 79, 81, 82 right, 83, 86 right, 87 left, 91 left, 92, 93, 95 left, 97, 100 both, 104 left, 105, 108 both, 111, 112, 113, 115, 117, 118, 119, 120 large, small left, small right, 122 both, 123, 125, 126, 127, 129, 130, 131, 134 both, 137, 138 left, 140 right, 141, 144, 145, 148, 149, 152, 153, 156, 157, 158, 159, 161, 163, 170 right, 171, 176, 177

Disclaimer

The information and recipes printed in this book are provided to the best of our knowledge and belief and from our own experience. However neither the author nor the publisher shall accept liability for any damage whatsoever which may arise directly or indirectly from the use of this book.

It is advisable not to serve dishes that contain raw eggs to very young children, pregnant women, elderly people, or to anyone weakened by serious illness. If in any doubt, consult your doctor. Be sure that all the eggs you use are as fresh as possible.

Original Title: *Beerenliebe. Genuss aus dem Garten und der freien Natur*

ISBN 978-3-86362-016-5

Project management, editing and proofreading: Julia Genazino

Layout and design: Christine Paxmann text · konzept · grafik, Munich

Translation from German: Katherine Taylor in association with First Edition Translations Ltd, Cambridge, UK

Editing: Sally Heavens in association with First Edition Translations Ltd, Cambridge, UK

Typesetting: The Write Idea in association with First Edition Translations Ltd, Cambridge, UK

Project management for h.f.ullmann publishing: Katharina Pferdmenges, Isabel Weiler

Overall responsibility for production: h.f.ullmann publishing GmbH, Potsdam, Germany

Printed in India, 2015

ISBN 978-3-8480-0806-3

10 9 8 7 6 5 4 3 2 1
X IX VIII VII VI V IV III II I

www.ullmann-publishing.com
newsletter@ullmann-publishing.com
facebook.com/ullmann.social